# Grassfed to Finish

## A production guide to
## Gourmet Grass-Finished Beef

# Grassfed to Finish

A production guide to

Gourmet Grass-Finished Beef

by

# Allan Nation

A division of Mississippi Valley Publishing, Corporation
Ridgeland, Mississippi, USA

First printing October 2005
Second printing April 2006
Third printing June 2008
Fourth printing August 2011
Fifth printing May 2016


Copyright © 2005 by Allan Nation

**Library of Congress Cataloging-in-Publication Data**

Nation, Allan.
    Grassfed to finish : a productrion guide to gourmet grass-finished beef /
by Allan Nation.
        p. cm.
    Includes index.
    ISBN 0-9721597-1-1
    1. Cattle--Economic aspects. 2. Pastures--Economic aspects. I. Title.

HD9433.A2N37 2005
338.1'76213--dc22

                                                            2005052535


Cover design by Steve Erickson, Madison, Mississippi

Manufactured in the United States of America
This book is printed on recycled paper.

# TABLE OF CONTENTS

> You cannot discover new oceans unless you have the courage to lose sight of the shore.
>
> Andre Gide

# INTRODUCTION

They say that in the life of every pioneering industry there is a moment when you know you have won. For me that point was hearing Jo Robinson's electrifying presentation on the health benefits of grassfed products at *The Stockman GrassFarmer's* Grassfed Meat and Milk Conference in Dallas, Texas in December 1999.

And, I was not the only one who saw the future that day. Today, you can trace almost every pioneer grassfed operation back to the 500 people in attendance at that conference.

In 1999, Jo Robinson and I could only find 40 grassfed direct-marketers in all of North America to list on her new grassfed consumer website www.eatwild.com. By early 2005, the number of producer web links on Jo's eatwild.com had grown to over 1000 and eatwild.com had recorded over one million individual visitors.

This rapid early growth was greatly accelerated by the BSE (Mad Cow) scare in late December 2003. This attracted national media and consumer attention to an industry that, to be frank, probably wasn't ready for it yet. The typical product being produced was, to be kind, uneven in quality.

While some grassfed beef was touted by the national media, as the "best beef they had ever eaten," a steak later in the year from the same producer using the same genetic base and grass could be dry and tough. I knew we could not build a viable industry on such inconsistency.

As you will read in this book, the United States has no history of ever having produced a "gourmet" grassfed beef product. As a result, there have been as many ideas about how to do this as there were producers. Since many pioneer grassfed producers could sell whatever they were producing for a premium price, they actually scoffed at the necessity for "gourmet" quality. One highly visible pioneer producer actually told a chef's trade journal that he would be "highly suspicious" if a tender, marbled beef product was truly grassfed.

What got me really depressed was how much worse our typical grassfed product was than what I had eaten in Brazil, Paraguay, El Salvador and Guatemala. These countries were hardly known for quality beef and yet the beef I ate in small towns and villages there was better than the average American product I was sampling.

Growing up in Mississippi, I had eaten home-raised grassfed beef until the late 1960s and never noticed it being significantly different from the Midwestern grainfed beef in the grocery store. What had changed in the subsequent 40 years? What where we doing wrong?

Riding to my rescue was a young rancher/researcher from Argentina, named Dr. Anibal Pordomingo. He showed me that what we were missing was the most basic knowledge of the "whole" necessary for producing a quality meat from grass. This whole consisted of easy-fattening cattle genetics, a high rate of gain prior to harvest and low-stress gathering, transport and harvest.

I first started going to Argentina in the late 1980s to get a sense of what a year-round grassfed beef industry at the same latitude as the United States' looked and functioned like. I had been to Ireland and New Zealand many times, but their mild maritime climates made their prototypes untransferable to the vast majority of North America. However, Argentina was different.

Here the winters were really cold and the summers

really hot. Drought and flood alternated just like in North America, and one could find the equivalent of Florida and Montana and every climate zone in between.

Eating their grassfed beef over a ten year period, I knew it was excellent year around, however, my grasp of the "whole" of what they were doing was severely hampered by my poor Spanish. I could see what they were doing but I couldn't see *why* they were doing it. All of their plowing, planting and planning sure seemed strange.

"There has to be an easier way," I thought.

I first met Anibal on a personal tour of the Pampas research station near Santa Rosa in 2000. I took note of Anibal's excellent command of English — he got his doctorate at New Mexico State — and when I brought a tour of American ranchers to the research station several years later I invited him to make a presentation on his research.

That presentation was on the difficulties in supplementing cattle on grass with grain.

Afterwards, we had lunch together and he asked what I had thought of his talk. I told him I was interested in producing beef **with no grain at all.**

He leaned back with a big smile on his face and said, "This is my passion as well. The only reason I did the grain talk is that is what most Americans want to hear."

I explained that there were a few Americans who wanted to hear the all-grass story and invited him to come to Tulsa, Oklahoma, and do a two-day seminar on a year-round forage chain. I chose Tulsa because it is located at the same latitude and rainfall as his home ranch in Santa Rosa. This seminar was a huge success and from that, a regular column in *The Stockman Grass Farmer* and

more seminars were initiated around the USA.

Another result was this book.

Without Anibal's input, I would have to tell you that a "gourmet" grassfed beef product is only possible in the late spring and early summer. With Anibal's input, I can tell you that *a "gourmet" grassfed product is not only possible year around, but can be produced virtually everywhere in North America where adequate moisture can be provided.*

Now I understand that this is going to require a huge suspension of disbelief on your part. All your life you have heard that grain is absolutely necessary for a quality meat. Now, you are going to hear this absolutely isn't so.

If you are like most ranchers, it will take you 5.5 years of rumination and study before you will be willing to start to put the guidelines given in this book into practice. During those years of indecision you will lose out on earning between $600 and $1000 a head that could have been yours by changing your mindset and your practices.

However, just by reading this book you will be ahead of the vast majority of ranchers who will refuse to read it. If I can just sow the seeds in your mind that there is another way to obtain  the North American consumers' dollar other than through the feedlot, I will have succeeded.

If not you, others will come, believe, learn, practice and prosper. We have already won. All that is left is to show you and 350 million other North Americans the inevitability of what will come.

Read on and discover why this is so.

# CHAPTER 1

# The "F" Word

For many years, ranchers have told me that a big factor holding back our circulation sales has been our publication's name — *The Stockman Grass Farmer*. Many beef producers have a deep emotional conflict with the idea of becoming farmers of grass rather than ranchers. Being a rancher is tied up in all sorts of deep group self-identification symbols from big hats, big belt buckles, and boots to the kind of pickup truck to drive. A rancher knows who he is because he has a long cultural tradition that he can identify with — one that has been highly reinforced and glamorized by Hollywood.

Ranchers are as American as apple pie. But what's a grass farmer? A hippie growing marijuana?

We quickly found out that T-shirts that read "Proud to Be a Grass Farmer" sold best to young people who self-identified with the latter. Grownup ranchers wouldn't be caught dead in anything with the "F" word on it.

Since our publication has always been sort of way out on the cutting edge of grazing technology, I always thought that the name served as a pretty good reader filter. If you couldn't get past the "F" word on the cover, you were unlikely to embrace any of the ideas espoused inside.

Hopefully, you will realize that a year around, gourmet, grass-finished beef product will require that its producers become "grass farmers" rather than ranchers.

An all-perennial, gourmet product will necessarily be a highly seasonal one. A lot of this cursing and philosophizing about the "F" word is really a cover for our lack of farming skills. For example, the dislike for tillage is as pronounced in New Zealand as well, which has no John Wayne ranching tradition or heritage.

However, for others who have no particular love for a romantic past, it is a "going back" to what they had just as gut-wrenchingly left a decade or so ago. Many have found that ranch consultant, Stan Parsons', admonition that a rancher's equipment inventory should consist of nothing but a wheelbarrow "and that only if you love machinery" works extremely well in the commodity cow-calf and seasonal stocker operations.

Indeed, one of the best ways to get return on capital up is to get the capital required down. Since the only profit-enhancing tool commodity producers have to work with is lowering the cost of production, getting your operation as closely in sync with Nature's cycle as possible greatly lowers cost of production. In other words, in a commodity-priced situation, management's goal is to purposefully strive for extreme seasonality. Coming to such a decision is not arrived at easily nor painlessly. And, neither is a reversal of this position once it has been achieved.

A good example of someone having made this painful "flip-flop" twice is Richard Parry who ranches in Ingacio, Colorado, which is near Durango in the Rocky Mountains. Parry credits stopping grain farming and haymaking with saving his ranch. He sold off all of his tractors and equipment at an on-farm auction a decade ago and has financially done well without them with an all-perennial, irrigated ranch. However, now he is back to farming again.

What changed?

The end product changed. A business truism is that if you change the end product you change everything.

Rather than selling commodity priced feeders, Parry

started direct-marketing his ranch's production as a "gourmet" grass-finished meat product. In Parry's case this product is lamb but his case is equally valid for beef producers. Here's what he told me:

"When we started our grassfed lamb-direct-marketing business, we were planning to continue lambing in the spring and market the lambs as grassfats in the late fall, just before the snow came. This seemed the logical, low-cost, modeled-after-Nature approach that we have been philosophically committed to following.

"As we became meat marketers, we learned that stores and restaurants do not close their doors in the winter; in fact, here in snow country the opposite is true. Colorado mountain ski towns like Durango and Telluride boom in the winter and they will pay a premium for gourmet, grassfed meats.

"In New Zealand's South Island, I saw that every sheep farmer set aside one third of his paddocks for farming winter grazing crops. They planted annual cereal crops for finishing lambs and brassicas for wintering ewes.

"Yes, the Kiwis were farming their way through the winter, but this would take a paradigm shift for me because of my philosophical aversion to the "F" word — farming."

Some of us have been noisy advocates for an all-perennial agriculture. Richard Parry had been one of those and yet he was one of the first ranchers in Colorado to adopt a "forage chain" of winter annuals that allows direct-grazed winter finishing even at his extreme elevation and winter cold.

"My cereal rye paddocks are as green and vegetative in January as they are in October. Gains have been respectable in January and February, which are our coldest months of the year. The lambs are brought in to a nearby scale on a weekly basis where the heaviest ones are sorted off for harvest.

"In the future we will target worn-out paddocks in need of renovation. They will receive soil tests and subsequent treatments for proper soil and nutrient balance. We will first chisel plow the paddocks to open up the irrigation hardpan followed by roto-tilling to destroy the old sod. After one or two years in annual crops they will be seeded back to improved perennial grasses and legumes.

"Each year we will move on to the next targeted paddocks eventually covering the whole grazing cell while we take both our soils and our pastures to high levels of quality grazing and production.

"This makes me a heretic to my 18 years of low-cost, low-input, modeled-after-Nature training, but you cannot consistently produce quality, grassfed meats under that school of thought" Richard said.

Amen.

Now, what brought about this paradigm shift for Parry? I think the key element is his statement "as we became meat marketers." Parry was no longer an anonymous rancher selling to faceless customers for whatever his production would bring. Now he was in charge of his product's price, its quality, and when and how it would be sold.

However, with freedom comes responsibility. *He was no longer solely responsible for satisfying his own needs. He now had customers with wants and needs. As a "meat marketer" he saw that his job was to satisfy them.*

This is the ultimate paradigm shift and one you will need to make to be truly successful in any kind of direct-marketing program.

# CHAPTER 2

# First, Some Background and History

A wise man once said, before you can hope to change things, you must understand why they are the way they are.

I found a book called *Traditional American Farming Techniques* by Frank D. Gardner that really helped me understand the "culture" of American agriculture and why grain feeding is so predominant in all species of North American agriculture.

This book was originally published in 1916 as *Successful Farming*. It was reprinted in 2001 by The Lyons Press. I bought my copy from a bookstore remainders table so it may be out of print again.

Agricultural economic historians have identified the period between the American Civil War and the onset of World War I as the longest time period of overall agricultural prosperity in American history even though it was a time of almost continuous deflation.

Prior to the Civil War most American farmers lost money. The availability of new cheap virgin lands farther West prevented the development of a national conservation ethic, and extensive, non-regenerative farming prevailed.

After the Civil War farm prices collapsed, the frontier closed and there was no more free land anywhere in the country. No longer could farmers obtain cheap virgin land, mine the nutrients out of it and move farther West.

For the first time American farmers had to learn to make the most of what they had where they were. They began to study the European soil building techniques they had previously been able to ignore. By focusing on the soil, American farmers began to make money even though their average yields per acre in 1914 were a third of Germany's.

This hefty 1100 page book is supposedly the repository of the cumulative farming wisdom compiled during that 50-year period of general farm prosperity. Of course, everything was "organic" then and the book could still be an excellent resource for today's organic farmers.

One of the things the book admits is that American knowledge about the care and feeding of pastures did not advance at the same rate as that of crops. This was particularly true of pastures for finishing beef cattle.

Now, here's a key point.

In Europe most of the food produced was consumed locally. This meant the farmer and the consumer knew each other, and information on the quality, or lack of, could easily be transmitted to the farmer. *If a farmer did an exceptional job word quickly spread and he was quickly rewarded for increased quality with a higher price and more customers. Consequently, quality as measured in flavor and eating quality became the major goal of most European farmers.*

Artisanal food production comes easy to European farmers because that is their heritage. It is still possible today to go up into the mountains in almost any European country and find a viable pre-industrial, grass-based agriculture still in existence that can be studied. In France, it is estimated that as much as 30% of their domestic food is still produced using "artisanal" production models.

However, the USA and Canada with their large land mass and widely dispersed cities created a different kind of agriculture than the more densely settled Europe. In North America, neither the farmer nor the consumer knew each other. It was a faceless relationship. They were separated by hundreds and perhaps thousands of miles and several layers of commission agents and middlemen. The only thing the North American farmer got paid for was meeting the minimum specifications.

If a farmer did an exceptional job, the middlemen got the extra profit — not the farmer. Consequently, cost control rather than product quality became the North American farmer's obsession. The only way the North American farmer could increase his profit was to lower his costs. His biggest cost was in getting his production from where he lived to where the consumer lived. As a result, reducing transportation costs became the primary way to increase profits.

A Canadian friend recently described his interior province's grain as a "freight-challenged commodity." However, this was even more true a century ago. Corn, the predominant American crop, was virtually worthless in the heartland where it was grown and had only slightly more value in the cities where it was seen as "pig feed."

Despite its low value, it was the most widely grown crop in the USA in 1916. At that time the USA produced three-quarters of the world's corn supply. This tropical import responded well to the long, hot summers in the heartland.

In 1916, the South grew almost as many acres of corn as the Midwest with Texas, Oklahoma and Kentucky being major producers. Rain in July and August was the most critical factor in determining corn yield. The Midwest had a more reliable rain fall in this period and became the corn belt. The average yield for the country as a whole was 62 bushels per acre.

In 1916, 80% of all the corn grown was consumed within the counties in which it was grown. Almost all corn fed to swine, beef cattle and sheep was home-grown. Dairymen were the only livestock farmers who tended to buy, rather than grow, their corn needs.

At this time, work stock consumed much of what a farm grew. A horse or mule ate approximately 70 bushels of corn or 100 bushels of oats per year. There was one horse or mule for every 30 developed acres in the USA.

Thanks to the transportation monopoly the railroads held, it cost far more to ship products across the country than today. Only high value products could pay the freight tariff for long-distance transport. Consequently, livestock production began to be used as a way to "densify" low value crops into something that could be economically shipped to the cities or to Europe where the highest price, or a price, could be gained.

In other words, livestock's primary purpose in North America was not to feed people but to add value to virtually worthless grain.

Here's how Gardner explained the 1916 economics of livestock production.

"It requires about 10 lbs of dry matter to produce one pound of beef or 30 lbs of dry matter to produce one pound of butter," Gardner wrote.

"The farmer in transforming such coarse products to a more refined one not only reaps the profit in the process of manufacture, but the pound of butter may be sent to a market a thousand miles away.

"One cent a pound for transporting butter would be but a small percentage of its value, but one cent a pound for transporting hay would be prohibitive.

"One bushel of corn will produce ten pounds of pork, but the ten pounds of pork can be shipped to market at considerably less cost than the bushel of corn."

So, our cultural history in North America is that of livestock's — including ruminant livestock's — primary role to add value to grain.

In 1916, chickens and hogs were the dominant non-work stock farm animals. Almost every farm had a few pigs and 50 to 100 chickens which were primarily kept as scavengers.

Swine were used as followers behind ruminants and would eat the cows' manure. They also utilized the skim milk and buttermilk from the many farms that made butter.

While grain was cheap and plentiful, protein wasn't. As a

result, leguminous pastures were very important as an inexpensive protein supplement to all species of livestock.

This leguminous pasture was also a critical part of providing soil nitrogen for the grain production.

"No cropping system will prove satisfactory for a long term of years that does not include at intervals of four to five years a leguminous crop such as clover, alfalfa or some of the annual legumes," Gardner wrote.

The two most common rotations were the traditional European four-course rotation of corn, oats, wheat and legume pasture and/or hay. In some cases, the legume pasture was lengthened to two years, making a five year rotation. In wet, warm-winter climates, this pasture period had to be extended to as many as five years due to the fast burnout of soil organic matter. Ruminant livestock were a way to add value to the necessary fertility building pasture portion of grain production.

"The cheapest way of feeding is to allow animals to harvest their own feed. The grazing of grasslands and the pasturing of cornstalk fields is typical of this process," Gardner said.

In 1916, four-fifths of all American farms reported having dairy cows. Cows grazed pasture in the green season and survived on corn stalks and straw in the winter. With such a diet seasonal production was the norm. At that time, the highest return to dairying was from farmstead cheese manufacturing and that is still true today.

Cheese manufacture allowed the cows to be in production only in the green season but provided a year around income from the sale of aged cheese. Cheese also was a dense high-value product that could be cost-effectively shipped to the cities. Cheese became the primary product of Wisconsin and Minnesota at the far Western edge of the dairy pasture zone because it had a high enough value to be shipped to the eastern cities.

In the South, dairying was practically non-existent as almost all dairy manufacturing requires a temperature of less than

58° F. This ambient temperature factor kept dairying concentrated is the summer-cool Great Lakes region until the advent of rural electrification. It was this lack of a high-value, ruminant product that made the rural South so much poorer than the Midwest.

Cheese's primary drawback was that it required a lot of skill and most farmers were loathe to invest the time learning the craft required. Farmstead buttermaking required less skill and specialized equipment than cheese and was the most widely made farmstead product in 1916 with two-thirds of all farms reporting butter manufacture.

Other farms separated the cream from the milk and sold this. A necessity of dairy farms producing cheese, butter or cream was having chickens and pigs to add value to the otherwise wasted by-products of whey, buttermilk and skimmed milk.

Male calves were typically vealed at six weeks of age or sold as milk-fat, four- to six-month-old "baby beeves." On many farms the farmer got the morning milk and the calf the evening milk.

Because low quality crop residues were the pre-dominant winter feed, the goal was to harvest the male animal before the end of the green season.

Many of the non-dairy beef cattle that lived largely in the South and Southwest were descendants of the original Criollo cattle brought to America by the Spanish and had a low-muscle phenotype similar to a dairy cow. In other words, the cattle were more hide and bone than muscle and were called "scrub cattle."

A major drawback to beef production in the hot summer in America was that even a small cow's beef was far larger than a single family could consume before it rotted. Consequently, moving these animals out of the rural areas to the cities where they could be quickly consumed became a necessity. This was initially done on the hoof and, after the railroads came, in ventilated stock cars.

These rural cattle were sent on consignment to the cities where commission agents would try to sell them to local meat packers. Often the cattle did not bring enough to pay for the rail freight. Consequently, beef production was not seen as a viable enterprise to most small farmers due to its high market risk.

Rural Americans' meat protein prior to electrification came primarily from small animals like chickens, wild rabbits and squirrels that could be consumed the day they were harvested. Winter-harvested pigs could be smoked and eaten year around but the only way to preserve beef was through salting. To de-salt the beef required a long period of soaking and was seldom completely successful.

What is frequently overlooked is that what created the great American beef industry was not the consumers' demand for more beef but industry's demand for more leather.

Until the advent of the small electric motor, factories relied upon a single stationary steam engine that transmitted its power throughout the building via line shafts and leather belting. As manufacturing grew rapidly after the Civil War, the demand for leather for belting pushed the price of cattle up in the East and started the great cattle drives in the West. Because leather was the primary product, little to no thought was given to meat quality. Most American fresh beef had to be tenderized with a wooden mallet to be edible.

Gardner wrote that there was a huge amount of con-sumer agitation in 1916 over the high price and low eating quality of the beef available. The majority of American beef at that time was still from dairy and "scrub" dairy-beef cross cattle. Needless to say, the heritage of American beef was definitely not "gourmet."

A major push at the time by Eastern gourmands was the promotion of the use of true English breed beef bulls on Ameri-can grade cows to increase their cutout yield and overall tender-ness. Here is the type of animal Gardner said a beef producer should be looking to produce for high quality meat.

"The essential characteristics of a good fattening steer

remain constant. A wide, strong, short head; short, thick neck; and deep, wide chest indicate constitution, and a deep roomy barrel indicates capacity.

"The type, quality, form and finish as indicated by the deep covering of muscle, even distribution of fat, high percentage of higher priced cuts of meats, high dressing percentage, smoothness and symmetry of carcass, and quality and texture of meat, are always associated with beef blood.

"A steer with a long, narrow head, long legs, or shallow body will not alter his type by fattening. A steer with a high tail or prominent hook-bones will finish into a fat steer with these same deficiencies.

"Cattle should be bred for early maturity otherwise they will grow rather than fatten and the cost of production will exceed their value. The majority of yearlings are marketed from 60 to 90 days before they are fat, which indicates that it is essential to secure calves of the type that will fatten easily."

These are all excellent observations and are still valid today for grass finishers. However, they largely fell on deaf ears. The problem then is the same problem as now. Few non-wealthy people are able to put in the time required for turning a herd around genetically.

Despite ample cow forage quality resources in 1916, Gardner said the primary constraint against owning breeding stock was the long production period it required. At that time, bank money was only available to livestock producers for periods of 90 or 180 days.

Gardner wrote, "It is possible for a farmer who has pasture to go to almost any bank and secure funds with which to purchase steers to consume his corn crop. It is impossible, however, for him to borrow the same money with breeding females for security, because three to five years must elapse before the increase will be marketable."

Another problem was that most farmers in 1916 were tenants rather than owners. Gardner said this limited them to short-time horizon activities like cash grains and cotton.

"The rapid growth of tenant farming has eliminated the production of meat from thousands of acres of land which should never have been plowed," Gardner wrote.

"The chief profit in cattle farming is the increase in the fertility of the soil. Where land is rented annually there is no incentive to build it up and increase crop production when a different renter may farm it the next year."

Grain feeding got an additional boost from the packing house industry. To increase industrial efficiencies, central beef packing plants were built at rail centers like Chicago and Kansas City. These large scale, beef only, plants required a steady year around supply of cattle to be efficient. Consequently, they began to promote the idea of grain feeding beef cattle. This message was reinforced with a higher out-of-season price.

By 1916, Gardner said the United States beef market was divided into a grassfed and a grainfed market by season. Cattle were largely killed off of grass from the middle of July to the first of December and came from feedlots the remainder of the year. However, there was no green season premium for grainfed cattle. The idea that grainfed animals were "superior" to grassfed animals was not widely promoted until the great grain glut that occurred after World War II due to the mechanization of field cultivation and the elimination of grain-guzzling horses and mules.

Gardner warned his readers considering the higher cost of grain-finishing to stick to the winter and spring months when the packers paid a seasonal premium. He said that the only way cattle would make money in the feedlot was if they had a forward margin on purchase-to-sale price.

"It is seldom that the value of gain in fattening cattle is equal to the cost of the feed consumed. Generally, a feeder weighing 1000 lbs can be purchased for one to three cents per pound less than he will bring when in prime condition and weighing 1200 to 1400 lbs.

"In the late summer and early fall the markets are usually

well supplied with beef that has been produced cheaply on grass with which the grain-fed cattle cannot compete profitably."

While calves were produced everywhere, large scale grass-finishing tended to concentrate in eastern Oklahoma and Kansas, southwestern Wisconsin and on upland mountain "balds" in the Appalachian states. These finishing areas featured either cool summers or naturally occurring highly mineralized grasses. Most also featured stony soils that couldn't be plowed.

He noted a major problem with grassfed cattle at the time was the lack of information on how to successfully finish cattle on pasture versus volumes of literature on how to do it with grain. This was particularly true in regard to the use of annual pasture, which could have helped to smooth out the annual flow of grass-finished beef and lessened the need for seasonal feedlots.

"While grass is the most important crop produced in the United States, there is not an important investigational project on the subject reported which the meat producer can use in a practical manner.

"More attention should be given to pastures to increase their carrying capacity by fertilizing them with manure or fertilizers, by thickening the stand of grass by natural or artificial means and by using grass silage during unfavorable periods.

"Waste by tramping may be prevented by restricting the area grazed by means of hurdles or temporary fences.

"Pastures of annuals require knowledge relative to the date crops must be sown to afford pasture when needed. In this respect it resembles the provision for soiling crops which are to be cut and fed from day to day.

"(In contrast to grass finishing) there is, however, so great an abundance of information as to methods of grain fattening that it is possible for those familiar with the publications and the general farm practices to recommend rations which are certain to produce rapid and economical gains in the feed lot with acceptable dressing percentages."

The dual-market of grainfed and grassfed cattle lasted in much of the South and West until the national supermarket chains with centralized buying programs replaced locally owned butcher shops.

However, an equally great impact was the shift from small-scale, on-farm feeding of one's own grain to huge industrial-scale, custom feedyards that made their money on yardage and a markup on the amount of grain fed. Prior to this, the time cattle spent on feed seldom exceeded 90 days because the cattle would become too fat.

However, these custom feedlots wanted animals that they could feed a high carbohydrate diet for five months without over-fattening. Because they quickly became the dominant buyer of feeder cattle, beef cattle genetics shifted from early-maturing, easy-fattening cattle to hard-to-fatten, late maturing, tough cattle. It is this shift that so bedevils grass finishers today and which will be discussed in the chapter on Genetics.

As radio commentator Paul Harvey says, "Now you know the rest of the story."

The reason so few of us know and respect the value of pasture is that a management-intensive grassland culture and the production of a direct marketed "gourmet" product is not a part of our cultural heritage the way it is in Europe.

We are not seeking a return to the past as some have claimed. We don't have an artisanal past. There are no old wise men we can go to for guidance. *You and I are pioneers of something that has never existed in North America.*

One of the things this lack of domestic history requires is that we be willing to look overseas for ideas and guidance. I know this is not in our heritage either and is something that will be resisted by 95% of North America's farmers and ranchers, and yet I have found no viable alternative.

Consequently, for over 20 years I have traveled the globe looking for pieces of grassland management information that could be put to work here to create, as they say in California, a truly "killer" product. This book is the result of those travels. Much of what you will read will have to be "tweaked" for your particular climate. I recommend you try all of these recommendations on a small scale first to give yourself the confidence that it can indeed be done in your area.

Everything in nature starts small and grows by multiplying previous successes. Your grass-finished beef business should do the same.

# CHAPTER 3

# Three Distinct Phases

Have you ever wondered why we have three distinct phases in beef production?

These three being cow-calf, stocker and finishing.

It's because each phase reflects a specific forage composition and quality required for that stage of production.

Too often, people choose an enterprise due to its reported profitability without fully understanding that it must match the quality of grass you have, or can grow.

## Cow-Calf

A mature beef cow can utilize relatively low quality forage resources for most of the year as long as her calving is timed so that she has time to create body fat reserves before calving. She can then supplement herself and her calf literally from the fat on her back during seasonal periods of low forage quality.

Backfat becomes butterfat in the calf's milk and bypasses the rumen where it provides a slow release protein and energy source. This is why calves on the cow can gain two pounds a day on grass where weaned calves will lose weight.

In the higher temperate latitudes, cows are beneficially affected by the extreme change in day length from winter to summer. This differential allows them to naturally cycle their calving toward the longest day of the year when forage will naturally be plentiful. However, the closer you get to the equator the less effective this natural cycle becomes because the days and nights tend to become closer in length.

For example, Biloxi has an hour more sunlight than Des Moines in the winter and an hour less in the summer. At the equator, days are the same length all year.

While the majority of the world's beef cows are found in hot, humid environments, individual productivity there is very low compared to the higher latitudes. In Florida, for example, the average reproduction rate is only 60% versus 85% in the higher latitudes.

The low reproductive percentage reflects both the low soil mineralization found in sandy, humid climates which lowers forage protein and digestibility and the general absence of managed seasonal calving.

Both of these can be overcome with management.

The hotter and more humid the summer is, the more critical it is that the cow be open during the dog days of summer. Otherwise, a syndrome described by South African animal scientist, Jan Bonsma, as "tropical degeneration" sets in.

Under this syndrome, calves are aborted, born deformed or unusually small. Most graziers miss noting the abortion and are perplexed when cows they thought were bred mysteriously end up open.

This problem is particularly acute in black-haired cattle and in regions where the air temperature and the humidity approach the cow's body temperature of 101°F and the night-

---

**Bermuda Grass Blues**
**Due to the lack of cool-season perennial grasses, Deep South grass finishers must rely heavily on both winter and summer annuals. This is extremely risky without supplemental irrigation and/or high quality stored forages such as alfalfa hay or annual ryegrass silage.**

time temperature does not fall below 60°F.

Thanks to generally cool nights, daytime heat is not as big a factor in semi-arid climates as it is in humid areas.

Whereas, cows in the higher latitudes tend to naturally move toward a June 21 calving date, this is not as pronounced in the Deep South due to declining day-length differentials and therefore must be managed for.

The lack of day-length stimulation requires cows to be in better body condition at calving in the South than in the North to stay on a 365-day calving interval. This naturally happens with late June calving.

Also, tropical forages such as bermuda and bahia peak in their annual forage growth on the longest day of the year. These same forages while green in March and April are low in dry matter and not growing over a snail's pace. They can literally starve an early spring lactating cow to death.

There is a saying in Mississippi that reflects this syndrome. It goes, "March shakes 'em and April takes 'em."

Another advantage of calving near the longest day of the year in the South is that it moves the breeding season to late September when daytime temperatures are likely to have moderated. This is particularly important for graziers using heat-sensitive, English-breed bulls.

All in all, *June 21 calving is much more critical in hot, humid climates than in cooler or semi-arid climates.*

In the mid-latitudes where cool-season grasses predominate, some graziers prefer to time their calving to mid-May to help control the spring flush. This works because cows double their forage intake upon calving.

Just keep in mind that a mid-May calving cow will require a bull that is willing to work in mid-August to stay on a tight calving schedule.

Do you have these? If not, you will need to calve later.

On infected fescue, fall calving (late August through October) can be a better economic option if one does a good job of stockpiling.

Grazing consultant, Jim Gerrish, said total cash out-of-pocket expense for a fall calving cow on well-rationed, stockpiled fescue in northern Missouri can be as low as $45 a year.

On non-infected fescue, late spring calving is better because it returns more to the land base because the cow's consumption is in sync with the seasonality of its production. This allows a higher stocking rate.

Remember, *a cow doubles her forage intake when she calves. Use this to balance your annual forage growth curve.*

## Stocker  Cattle

Stocker calves are weaned calves less than a year in age. These can range from very young 250 pounders to near-yearling 650 pounders but are typically around 450 to 500 lbs at weaning.

By using fence-line weaning whereby the calves can see and hear their mothers, the weight loss and sickness that accompanies other forms of weaning can be avoided.

Due to their small rumens, stocker cattle need a highly digestible grass that can pass through them quickly to gain well. The lighter the calf is in weight the higher in digestibility the forage must be.

This can be managed for with Management-intensive Grazing (MiG) by giving young, light calves a high degree of selectivity (low stocking rate) and following them with an older class of animal to eat the remaining sward.

Putting light, fresh-weaned calves onto an all-hay or silage diet normally produces gains of less than a pound a day.

The two best grasses for newly weaned calves are highly digestible winter annuals like oats, rye, triticale or annual ryegrass in the late fall or spring perennial pasture.

---

**The best time of the year to feed high quality annual ryegrass silage is during the summer to increase the average daily gains of your stocker herd.**

On winter annuals, weaned calves can gain in excess of 1.5 lbs per day during deep winter and double that rate in the spring.

Winter annuals can be grown virtually everywhere in the continental United States and provide excellent winter stocker calf gains at a very low cost.

If you do not have winter annuals, it is best to leave the calf on the cow through the winter and wean the following spring.

Yes, the cow only produces a teacup full of milk at this time, but this milk bypasses the rumen and can allow the calf to continue to gain over a pound a day on pretty rough feeds.

In the Argentine winter of 2005, Anibal said his calves were still on their mothers at nine months of age and weighed 580 lbs. Purchased stocker calves of the same age but weaned at five months of age only weighed 360 lbs.

Anibal winters his cows and calves on corn stubble. A pregnancy exam showed a 97% breed back.

His previous year's calves, which were also late-weaned at nine months, averaged 1.8 lbs per day on their mother, 1.6 during their stocker phase and 1.98 during their four-month-long finishing graze. The steers were harvested at 1035 lbs at 19 months of age.

"The key thing for a 19-month harvest is the late weaning," he said.

"In a different cow-calf program (earlier weaning), we would be four to five months from finish at 19 months of age."

This long lactation will draw down the cow but with a late May or late June calving, she will have plenty of time to recover before calving.

Actually, it appears that getting thin in winter and recov-

---

**85% of the entire gain from cool-season perennial pastures is made in the first 90 days after initial spring green up.**

ering in spring produces a faster breed-back than artificially maintaining the cow in good condition all winter.

With the early-maturing, easy-fattening, smaller cows, leaving the calf on the cow for up to 10 months is frequently necessary to prevent the cow from becoming too fat prior to calving.

*Please, note that this long lactation is part and parcel with June calving. Don't try this with early spring calving cows because they won't have time to recover their body condition before calving.*

If you have digestible, high-protein pasture, the stocker phase has the widest production margin in beef production because so much of the animal's weight gain is water.

The stocker grazier is also riding the adolescent growth curve, which can be very high with male animals.

Because lean flesh is protein and stocker cattle are growing only flesh and bone, they respond well to high protein forages and their gains are typically not negatively affected by the use of artificial nitrogen fertilizers on cool-season forages. Such is not the case with finishing weight animals, as I will discuss momentarily.

*The majority of production mistakes with stocker cattle are made by not respecting the seasonality of the digestibility of their forages.*

Most graziers think the reason stocker cattle's gains are slow in the summer is because of the effect of the heat on the cattle. It's not. Stocker cattle have gained over three pounds a day at Mississippi State in August on grazed green corn.

*The big problem is the heat's effect on the grass.* Heat above 86°F makes the grass much less digestible.

---

**Weigh your stocker and finishing cattle every 90 days. Sell all stocker cattle gaining less than 1.3 lbs per day.**

---

Here's the result:

With cool-season perennials, 85% of the stocker animal's green season gain is made in the first 90 days of the grazing season. This means if you typically start grazing on May 1, 85% of the gain you are going to get for the whole green season will have been gotten by August 1.

With warm-season perennials, this 85% figure will be reached in only 60 days or around the 4th of July.

Anibal said too often graziers get so fixated on the finishing cattle that they ignore their stocker cattle. This tends to particularly be a problem in the late summer and fall when the grass is green and yet gains are collapsing. One thing that can help temper this seasonal quality decline is to add legumes.

Vetch added to winter annuals can increase late spring stocker gains by as much as a pound a day. White clover can do the same thing with cool-season perennials.

White clover is unique in that it is the only perennial forage whose digestibility is unaffected by the heat. However, it requires an annual rainfall of around 40 inches (or equivalent irrigation).

Another gain enhancer is crabgrass. Jim Gerrish was able to maintain a complementary stand of crabgrass in his fescue in Missouri with just grazing management. This helped keep his mid-summer stocker gains in the acceptable 1.5 lb range. Crabgrass, a highly digestible warm-season annual, is less affected by the heat than a perennial cool- or warm-season grass.

Grazed alfalfa can produce a pound a day in August, which is better than a half pound from fescue or bermuda, but is far inferior to white clover for mid-summer animal gains.

---

**For best gains while stretching available grass supply, feed stored forages like hay or silage in the morning and let the cattle graze in the afternoon when carbohydrates are highest.**

Another gain enhancing trick is to feed spring-cut annual ryegrass silage or spring-cut alfalfa hay as a supplement to heat-stressed perennials. Anibal said that due to grass silage's low dry matter it was better used as a summer digestibility supplement than as a winter supplement to winter annuals. Both ryegrass silage and spring-cut alfalfa hay will be more digestible than perennial pastures during hot weather and will help the stocker cattle stay in the 1.5 lb per day range.

Once the weather turns cool, we enter another low gain period. The regrowth on cool-season perennial pastures following the summer's heat is very low in dry matter and will produce unacceptably low gains unless supplemented with hay.        Allowing stocker cattle free-choice access to alfalfa hay in the fall can double stocker gains on winter annuals to 1.5 lbs per day.

Once frost hits the cool-season annuals they will naturally increase in dry matter, and such dry matter supplementation will no longer be necessary. However, *the best Argentine grass-finishers keep free-choice alfalfa hay available for as much as 10 months of the year* for dry matter supplementation and to avoid accidentally shorting the finishing cattle.

Anibal said that while an all-legume hay should be your primary stored forage in a stocker-finishing program, feeding the hay or silage of the next forage the cattle will be going to for two weeks prior to making the shift will help prepare the rumen to digest it.

This is not necessary when shifting from cool-season to warm-season forages but is necessary when shifting from cool-season perennials to cool-season annuals or from warm-season perennials and annuals to cool-season annuals.

---

**Grass-finished beef starts with weighing your cows. This will give you a rough idea of what weight your cattle will finish. Typically steers will finish about 100 lbs heavier than their mothers, and heifers will finish about 100 lbs lighter.**

# Grass Finishing Cattle

Most American graziers, myself included, initially considered finishing weight cattle as just heavy stocker cattle.

However, such is not the case.

They are an entirely different class of livestock.

Anibal said *the big difference between the two is that the stocker cattle's weight gain is flesh and bone; whereas, the majority of the finishing animal's weight gain is fat.*

You can grow flesh with protein but you need energy (carbohydrates) to produce fat. *As a result, highly proteinaceous forages with low energy levels that can produce excellent stocker gains are a total bust with finishing animals.*

"The ideal finishing forage would be one that is roughly balanced between energy and protein," Pordomingo said.

Another example of production differences is that the use of artificial nitrogen on cool-season forages raises the protein content of the forage and creates a protein/energy imbalance is a production no-no with finishing animals. (This is not a problem with warm-season forages where energy is almost always adequate.)

Many American forage research and extension people have missed this key point by directly interpolating stocker cattle gains directly as what could be achieved with finishing cattle. This has been particularly true when discussing the fall low gain syndrome.

*The best finishing gains will come from forages grown with slow-release, natural soil nitrogen.* As such, finishing cattle are perhaps as easy to produce organically as not.

---

**Nitrogen fertilizer has little place in a grass-finishing program as it artificially raises the protein of the grasses.**

**With finishing cattle, we would like proteins and carbohydrates to be roughly in balance.**

---

Medium-frame cattle must gain over 1.7 pounds per day to produce marbling fat. At gains below that threshold, body maintenance consumes the majority of the energy.

Such finishing level gains are not possible from perennial pastures with fattening cattle in the fall. Here's why.

As the daylength shortens, perennial plants start translocating their energy reserves from the leaves to the plant's roots. This creates a high protein/low energy plant, which is great for stocker cattle but totally unacceptable for fattening cattle.

In the spring, the reverse occurs. This is why perennial forages that can finish cattle easily in the spring can't in the fall.

It was once thought that you could supplement these cattle with energy by feeding very small amounts of grain or molasses and balance the protein. Good results were found when these carbohydrate supplements were kept to less than five pounds per day. However, higher levels of supplementation resulted in a rumen microbe shift and a drop in the animal's ability to digest forage.

As a result, on-pasture, carbohydrate supplementation just as frequently resulted in the animal's average daily gain going down as up.

"We (Argentine research and extension) now recommend that ruminant animals should only be supplemented with forages," Anibal said.

He said graziers desiring a fall-finished animal were caught between a rock and a hard place. Your perennial forages will be too low in energy to create fat and young winter annuals will be too low in dry matter to create marbling level gains.

---

**In the humid mid-latitudes (above 35" of rain), a beef finishing ranch would typically have 70% of its land area planted to cool-season perennials, 20% to winter annuals, and 10% to summer annuals. Due to needed soil digestion periods following plowdowns, not all of the land is covered at all times.**

This results in two options for graziers.

One is to skip trying to finish beef animals in October and November.

This is the option most of the grass-finished world has taken. These cattle will finish easily on frost-hardened winter annuals in December or on spring pasture the following year.

The only other option is to use a late-planted, direct-grazed, greenleaf corn.

Of course, the really big problem is the mental paradigm shift required. We have not known nor fully appreciated the nuances of the seasonality of grass and have culturally conditioned ourselves toward fall being the harvest season.

A June-born, medium-frame steer can be easily finished on perennial pastures to gourmet quality by its second birthday. *Once you master this base level you can start to look at finishing cattle on a more year-round basis.*

# CHAPTER 4

# Theory of Finishing

A rule of thumb in business is that if you change the desired end product you change everything. As you saw in Chapter 2, the primary role of animal agriculture in North America throughout our history has been to add value to grain. As a result, our animal genetics and management emphasis have been geared toward a grain-finished end product rather than a grass-finished one.

*With grain feeding, the primary problem is the animal getting too fat. With grass, it is getting the animal fat. The animal genetics designed for one system will not be the best fit for the other.* However, we must do what we can with what we've got.

I will cover grass-finished genetics more in depth later. At this point I will just say that animals selected for gourmet grass finishing should be medium-framed English breed cattle. As I will explain, the closer you can get your cows to 1000 lbs, the easier grass finishing will become. Because product consistency is a highly valued trait in grass-finished beef, crossbreeding is not widely practiced nor recommended, and some graziers use linebred bulls to strengthen the eating quality predictability of their beef even more.

The most important thing to realize early on is that *the protocol for grass finishing is created by the need to produce an end-product that is very similar, if hopefully not superior to, the grain-finished product to which most North Americans have become accustomed.*

This means that for the North American consumer it should be both tender and marbled. (In a later chapter, I will go into some of the ways marbling improves the beef eating experience.)

Marbling is not highly prized in France, Argentina or New Zealand, but it is in North America. Again, it is the end product that determines what is possible. In America, we tend to cook our steaks hot and fast. In France, they think nothing of slow cooking their beef for four to five hours.

An Argentine barbecue grill, called a *parilla*, is designed so as to give the chef excellent control over the temperature of the fire to keep it from getting too hot. As in France, low heat and slow cooking predominates. Because their cooking methods compensate somewhat for beef tenderness that is somewhat less tender than in North America, both countries tend to concentrate more on beef flavor. Nearly all overseas people accustomed to eating grassfed beef negatively comment on the lack of flavor in American grainfed beef.

The point is that the way the world looks depends upon where you stand. We all like what we are accustomed to the best. The more we can get our grass-finished beef to match the eating attributes of grain-finished beef the more customers we will be able to attract.

I do not deny that we have been able to sell a goodly amount of poorly finished grassfed beef in North America for a premium price to very health-conscious consumers, but my dream is not to equal, *but surpass*, the eating quality of grain-finished beef. When all the stars have been in alignment we have been able to do this, but too often we didn't know what we were doing right and consequently couldn't replicate it.

What I hope to do with this book is to teach a production protocol that will produce a "gourmet" beef product every time.

But, first, let's take a closer look at the basic theory of grass finishing.

In grass-finished countries, the finished class of animals is the highest priced. It requires this price premium to offset the fact that stocker cattle, due to their lower body maintenance resulting from their smaller size, produce more beef gain per acre. This price premium also helps pay for the much higher degree of skill finishing requires. As the old saying goes, it requires "The eye of the master to fatteneth the cattle."

When cattle — start to finish — fatten, things start to get very difficult from a grass standpoint, because *fat requires the forage to be highly digestible and rich in soluble carbohydrates.* This is frequently called sugar. These soluble carbohydrates are not only determined by the plant species but stage of growth and the level of soil mineralization. For example, the level of soil calcium is directly correlated to the soluble carbohydrate level of the grass.

For finishing, the ideal protein/soluable carbohydrate ratio would be one-to-one. This ratio can be greatly upset

---

### Producers Not the Best Judges
### for Their Own Meat

**In a blind taste test, Jo Robison, author of *Pasture Perfect* and chefs from the Seattle area sampled eleven frozen ribeye steaks from grassfed producers in eight states. The results indicated a huge variety in meat quality.**

**The number one ranked grassfed steak surpassed all others in all categories — tenderness, juiciness and flavor. It scored 52.5 out of a possible 60 points. The second ranked steak was 5 points lower than the other grassfed steaks. Overall scores ranged from 52.5 to 28.**

**Only the producer of the winning steak said prior to the judging he felt he needed to improve. All other producers said they were satisfied with their steaks' tenderness and flavor.**

through the use of artificial nitrogen fertilizers. As a result, *finishing pastures are traditionally high in legumes, not only because of their higher digestibility, but because the soil nitrogen they create is in a slow release form that does not cause the plant proteins to spike as artificial nitrogen does.*

With finishing weight cattle, artificial nitrogen can frequently cause daily gains to fall in half.

A high protein/low carbohydrate forage that traditionally produces over two pounds a day with stocker cattle will frequently produce an average daily gain half that amount with finishing cattle. *Any weight gain in a fully grown animal must come primarily from the creation of fat.* This requirement for more carbohydrates is compounded by the fact that the animal is now heavy and has a much higher body maintenance requirement than a stocker calf.

Fully 75% of the daily energy intake the animal gets will be consumed in body maintenance. *The slower the animal gains the larger the proportion of daily intake that goes to maintenance. This is why the animal must gain in excess of 1.7 lbs per day to exceed this maintenance threshold.* Below this threshold no intramuscular fat is created.

To cross this threshold requires that we use a forage that is at least 65% digestible, at least 20% dry matter, with a protein level range that does not exceed 18% and a soluble carbohydrate content of at least 15%. With the exception of the mid-spring to early summer period, this is an "unnaturally" high quality forage. It can be created at other times of the year, but it will not be "natural" and will require the use of annual forages.

*This lack of understanding of the need for forage carbohydrates is a core reason for the poor reputation for grassfed beef in North America. Too many North American graziers think grass is grass and do not appreciate its subtleties and distinctions.*

While a "gourmet" beef product is a whole of many factors — including age at harvest, the presence of marbling, the genetic phenotype used, low stress handling and the rate of gain — most of us have considered grass-finishing to be an extension of stocker-growing when in fact it is a very different— and a much more difficult — enterprise.

The end result has been that most of the grassfed beeves currently being sold in North America would be more properly defined as "heavy feeder cattle" rather than "finished cattle." My definition of "finished" is an animal that will grade USDA High Select or Choice. In other words, they have a good degree of intra-muscular marbling.

*To produce intra-muscular fat requires an average daily gain in excess of 1.7 lbs per day for 60 to 90 days prior to harvest.* Range and tame warm-season grasses just barely past this minimum threshold and only in the late spring.

In most of the United States, cool-season perennial grasses will drop below this minimum threshold in late July or early August even with irrigation. Even with the more tolerant stocker cattle, 80% of the entire season's gains occur in the first 90 days after initial spring green up. This is due to daytime temperatures being in excess of 87°F which lowers the digestibility of all perennial grasses and most legumes including alfalfa and red clover.

In the Deep South, daytime temperatures will reach 87°F for 120 to 150 days a year. While this is the most extreme region, in almost no region of the Continental United States is there a total absence of such hot temperatures.

It is common for weight gains to drop 20% in early summer and 50% by late summer when compared to mid to late spring gains. While most graziers are willing to concede low summer gains, most thought their fall gains would rebound to levels similar to mid-spring.

However, *due to the shortening of the day length, the perennial grasses start to shift carbohydrates from their leaves to the roots to prepare the plant for over-wintering. This drops the carbohydrate percentage in the pasture too low, and marbling potential drops too.*

This problem is compounded by the fact that this fall regrowth from the summer slump is low in dry matter just like the early regrowth in the spring.

However, because the plant is growing slower, this fall "washy" stage lasts much longer — often until the grass is "hardened" by a frost. Gains in the two weeks following first frost will often double as the dry matter percentage increases.

The problem then becomes that there is not enough time for the animal to marble before the perennials go dormant for the winter. Remember, *we need 60 to 90 days of high gains to finish an animal.*

This problem of low dry matter in both spring and fall can be greatly helped by the feeding of alfalfa hay free choice. This will often double average daily gains. Many beef finishers keep alfalfa hay available free-choice year around as an insurance policy against dry matter deficiencies and grazing management mistakes.

The combined whammy of the "summer slump" and the "fall syndrome" means that a high-quality, marbled-fat, grass-finished product can only be produced "naturally" with perennial grasses in the late spring to mid-summer period. Because *fully grown animals once fattened tend to stay fat as long as they continue to gain*, the harvest period can be lengthened considerably beyond this period.

---

**Production Skills Needed for Gourmet Grassfed Beef**
1. **Knowledge of soils and plant species.**
2. **Knowledge of how to plan forage sequences and transitions.**
3. **Knowledge of how to practice leader-follower, Management-intensive Grazing.**
4. **Knowledge of how to part, sort and load cattle quietly.**
5. **Knowledge of how to interpolate outside visible fat cover with eating quality.**

The fact that beeves harvested in early September might be marbled does not mean they marbled in August. This has confused many graziers about the true quality situation in their pastures.

The problem is that *animals that were still growing during the late spring/early summer "finishing" window will not subsequently fatten on late summer and fall perennials. This makes the production of grass-finished beef from perennial forages extremely seasonal.*

Some grass beef finishers have recognized this limitation and accepted it. Many of these graziers sell their beef frozen and through seasonal marketing venues like farmers' markets. As a result, seasonal finishing fits their market well.

New Zealand's "table grade" beef production is largely seasonal and confined to the late spring to mid-summer. You will read details of this in the chapter on Proven Prototypes.

The real problem has come from graziers who have tried to fall-finish animals that were growing during the natural finishing period or on hay during the winter. Typically, these are graziers who do not wish to over-winter heavy animals for a second winter and harvest them at the end of the green season with no regard to their degree of "finish."

Because meat tenderness is highly correlated with rate of

---

**How to Lower Risk**

- Avoid overstocking at all costs.
- When animals are finished, move them promptly to the abattoir.
- Do not use restricted diet/compensatory gain programs.
- Plan to have access to both annuals and perennials during high weather risk periods.
- Buy high quality alfalfa hay or make high quality legume/grass silage.

**Dr. Anibal Pordomingo**

gain prior to slaughter and these animals were typically gaining a pound a day or less, this has resulted in some pretty awful eating stuff even when aged for 14 to 21 days.

Nothing has confused consumers more than the huge seasonal variation in the quality of North American grassfed beef. The beef they buy in mid-summer is wonderful but the beef they buy in the fall is awful.

I am far more interested in eliminating this huge quality differential in product than I am in promoting year around grass finishing. In fact, I frequently warn new consumers to buy their grassfed beef before mid-July.

In the course of this book, we will address this seasonality problem, but first let's start where you will probably start, which is on a perennial base forage.

## Where to Start with Grass-Finished Beef

With the exception of Eastern gammagrass, warm-season perennials have virtually no place in a grass finishing program. In fact, unless you have a very easy fattening breed, tame, warm-season perennials such as bermudagrass will need a predominant companion legume to achieve a minimum 45 day finishing rate of gain. In most of the lower South, this finishing period will occur from mid-April until the first week in June. In years with cool and wet late springs, this period can extend until the fourth of July.

Eastern gammagrass is a genetic first cousin to corn and is unique in that it is the only warm-season perennial capable of producing finishing gains through the heat of the summer. This is a very difficult and slow grass to get established, but if you are in the Deep or Middle South it would pay to consider it.

The key point here is that *with grass-finished animals, heat is a bigger a problem than cold.* The only solution to both is the planting and use of annual forages. I will cover this in the chapter on The Forage Chain.

A high quality beef eating experience starts when the animal is born.

Because calves are born not fully developed *it is critical that the animal have enough milk to be fully fed in the first 13 days of life.* Calves that do not get enough to eat at this time become what are called "dogies" and will not grow and fatten adequately no matter how well they are fed later in life.

Tucson, Arizona, beef nutritionist, Dick Diven, said the best way to insure this happening is for the animal to be born at a time when its mother has had time to fatten prior to calving.

The most common calving season is late spring to early summer. In hot humid regions, particularly with black cows, it is very important that the cow be open during the worst of the summer's heat. Otherwise, the cow will frequently abort the calf or it can be born dead or with serious deformities. Animal scientist, Jan Bonsma, called this syndrome "tropical deterioration."

Avoiding these problems is accomplished by *calving as near as possible to the longest day of the year or June 21st.* This has the added benefit of moving the breeding season to late September when summer temperatures have moderated. Also, in the South, June 21st is the peak of the annual growth cycle of warm-season grasses.

In the cooler upper latitudes, June 21st is also good in that it is when the grass growth typically becomes surplus but the cow has had time to fatten prior to calving. In the mid-latitudes, calving typically occurs a month to six weeks earlier.

In all three of these regions, by timing calving to the late spring, the cow can be used as an effective tool to keep your finishing pastures vegetative, eliminating the need for mechanical pasture topping and/or hay or silage harvest.

Diven said late spring calving has another benefit as well. He said *the second "critical moment" in a beeve's life is when it is 65 to 70% of its mature weight (700 to 800 lbs). This weight is typically reached one year after calving.*

He said beeves that are gaining weight at this time will form the intra-muscular fat cells that will be filled with fat during the subsequent finishing. Animals who are losing weight at this time will not form these fat cells but will instead develop connective tissue and gristle. As a result, these animals will always eat tough even if properly finished later.

*By calving during the spring lush, these subsequent yearlings are guaranteed to be gaining weight the following year.* As a result, they will subsequently marble well and eat tender when harvested the following year.

It is during this pre-weaning period that the size of the loin eye is determined. *The better the nutrition the calf is receiving the larger this valuable cut will become.*

September and October fall calving is often practiced in the Fescue Belt as it allows the use of low-cost, leased, endophyte-infected, fescue pastures. Through the use of late-summer and early-fall fescue stockpiling, many graziers have found fall calving can be nearly as economical as spring calving in this region on a cash out-of-pocket basis.

*Endophyte-infected fescue should never be used in the final finishing phase due to the highly objectionable off-flavor it produces in the meat.*

Because so much of the eating quality of the animal is determined prior to the animal reaching finishing weight, most gourmet-oriented, grass finishers are loathe to buy-in calves and yearlings from unknown producers. In Argentina, grass finishers

---

**Getting Started in Grassfed Beef**
**1. Weigh your cows.**
**2. Add 100 pounds to this weight. This will be the target slaughter weight for your steers.**
**3. Divide this weight by 1.4. This is a good average daily gain for a 365-day year. This will give you the number of days from birth until slaughter. For example, if your target weight is 1200 lbs, this will require 857 days or 28 months. If your target weight is substantially more than 1200 lbs, change your herd's genetics before starting.**

typically will have a separate cow-calf ranch in a lower quality grass region to produce the calves for their operations. However, they emphasized to me that having some cows were useful on even the most fully developed finishing ranch.

The Argentine rule of thumb is that a fully developed finishing ranch needs about 20% of its total animal weight in cows for grass quality control purposes. Because cows virtually double their intake of grass at calving, timing calving to when the grass growth rate is surplus to what your stocker and finishing animals can consume is the best choice for most graziers.

Now, here's the true beauty of beef cows.

Beef cows cushion their calves from the seasonal drop in forage quality through the butterfat in their milk. As a result, *a calf on the cow can gain two pounds a day on a grass that would only produce a half pound a day on a weanling.* This allows the grass finisher to use the cow as a tool to condition the pastures for the stocker and finishing cattle.

Keep in mind, that in cow milk it is the amount of milk fat that is important — and not the milk quantity. Cows in very late lactation can still produce enough fat to keep the calf gaining well even though the quantity of milk will be quite low. *Short-weaning has no place in a gourmet grass-finishing program.*

For the best eating beef, it is important that the animal not be weaned onto a forage that will not produce a similar on-the-teat average daily gain. The two most common forages for this would be winter annuals two weeks after the first frost in the fall and spring perennials three to four weeks after initiating growth.

The key element with both of these forages is to *delay weaning until the dry matter has risen to the point where good gains are possible.*

Allowing your beeves access to free-choice alfalfa hay during the low dry matter periods in the spring and fall can double average daily gains.

California grazier, Mac Magruder does this and says that

his hay consumption averages about six pounds per day.

It is far better to bring the calf through the winter and difficult late summer period on the cow.

Research has shown that cows that lose a significant amount of weight during the winter but then have time to fatten before calving actually are more fertile than cows that are over-wintered too fat.

If you are buying in calves, always select those that are showing good condition as this indicates a genetic propensity to fatten. *Never buy thin calves and do not try to finish for gourmet beef any of your own home-raised calves that are significantly thinner than their herd mates.*

Ironically, the fleshy calves that will eat the best are normally discounted at the auction as most stocker graziers are seeking compensatory gain.

Male calves should be castrated by seven months of age. Late castrated males produce a less tender, strong-flavor meat. Un-castrated males have the same problems plus will not marble regardless of the grass' quality. Spaying heifers at wean-ing will make them fatten faster, calmer and easier to manage.

If you are coming to grass finishing from the commodity stocker business, it is important that you forget most of what you learned. *The use of compensatory gain has no place in a gourmet program.* For the tenderest meat, the animal should never be allowed to lose weight at any time in its life.

---

**Gourmet Grassfed Beef**

- ■ USDA Select or Choice
- ■ Aged minimum of 14 days.
- ■ Recognized genetically tender breed.
- ■ Less than 36 months of age.
- ■ All grass (no hay or grain).
- ■ No artificial hormones or antibiotics.
- ■ Taste and tenderness tested before sale.
- ■ Tougher cuts mechanically tenderized.

Animals that have had shipping fever or any other respiratory disease that produced a high fever should be culled. Argentine research has shown that the meat from these calves will not retain moisture when cooked and will be a poor eating experience.

I am assuming you have a reasonable level of competence with rotational grazing and pasture subdivision. If not, get this knowledge before trying to grass finish any animals. An excellent book on this is *Management-intensive Grazing* by Jim Gerrish.

With finishing animals we want them to have a high degree of grazing selectivity and do not want them to take over 50% of the grass on offer. The primary purpose of pasture subdivision is to utilize this leftover grass with another class of animals. This is called leader-follower grazing. Without this second grazing, grass finishing is necessarily highly wasteful.

This second grazer can be yearlings if your grazing management is precise and timely, but most graziers feel more comfortable with cows and calves.

Anibal Pordomingo said there is no research showing any increase in eating quality by having an animal gain in excess of two pounds a day. However, he has lots of research showing that gains of less than 1.3 lbs per day in the stocker portion had a tremendous impact.

"You will do more for eating quality by eliminating the slow gaining tail end of your stocker herd than you will in trying to increase an already good rate of gain in your finishing herd," he said.

> **Contract grazier, Frank Holmes of Tylertown, Mississippi, grazed 40 head of English breed steers on annual ryegrass during the 2003-2004 winter for the New York Beef Company. Steers averaged 3 lbs per day and graded 40% USDA Choice, and the remainder High Select when harvested off of grass in early May.**

He recommends that all your cattle be weighed every 90 days and all animals that are gaining less than 1.3 lbs be culled and sold at auction rather than taken on to grass finish.

"An ideal finishing program would be one that produced 1.5 lbs per day for the first six months after weaning and two pounds a day for the final six months. If an animal is gaining at a steady rate, it is possible to fatten it before it is physiologically mature. This allows for a long harvest window. We can harvest the animal at 900 lbs or 1100 lbs. Such a long harvest window is the key component of a year-round, grass- finished beef program," he said.

Unlike grain-finished beeves, on grass you cannot over-fatten them but it is easy to under-fatten them.

### Start by Weighing Your Cows

The general rule of thumb is that a steer will not finish until it weighs 100 lbs more than his mother and a heifer will not finish until she weighs 100 lbs less than her mother. This is the "physiological maturity" to which Anibal was referring.

Two years is 730 days. If an animal gains 1.3 lbs per day (the Argentine and New Zealand average) it will weigh 949 lbs on its second birthday. If you are really good and it gains 1.5 lbs, it will weigh 1095. Now if your cows weigh 1400 lbs, you better have them gain two pounds a day every day of their life and you should probably only try to finish the heifers.

This is why *all grass finishing programs should start with weighing your cows.* Without this critical number you will not have a target for your finished weight.

An ideal cow size for grass finishing is around 950 lbs. Believe it or not this was the USA average cow weight 30 years ago but you will have a hard time finding one today.

So, what would we lose by downsizing our cow size?

Dr. Allen Williams asked himself this when he was teaching at Mississippi State. He is now a branded beef consultant with the Jacobs Alliance and works with several grass finished beef producers.

He analyzed Integrated Resource Management compiled data from several Southeastern states for the years 1997 to 2001 and came to the conclusion that rather than losing ground financially, cow-calf producers would actually be gaining by going back to smaller cows.

He said for cow productivity to have any financial meaning to ranchers it has to be translated back to a per acre basis. This is because the available land base is the primary economic constraint in ranching.

Williams said if you take the mid-latitude average of four acres of fescue per cow on a year around basis, a 400-acre ranch will support one hundred 1000-lb cows, eighty-four 1200-lb cows and only sixty-seven 1500-lb cows.

The more efficient 1000-lb cow can be run for $69 a year less per head than the 1500-lb cow based upon consuming 2.5% of her body weight per day and dry matter costed in at 15 cents per pound.

Also, the 1000-lb cow was the only weight category that weaned 50% of its body weight at 510 lbs. In contrast, the 1500-lb cow weaned only 41% of its body weight at 610 lbs.

The 1000-lb cow also had the highest reproductive rate at 87%. This was a full ten percent better than the 1500-lb cow at 77%.

This higher reproductive rate means the one hundred 1000-lb cows produced 87 calves versus only 52 from the 1500-lb calves.

Even though the 1000-lb calves were 100 pounds lighter (510 @ $82.50) than those from the 1500-lb cows (610 @ $80.00) they produced $11,229 more total income from the 400 acres because there were more of them.

So, **with the smaller cow you have a cow that is both cheaper to operate and that actually produces more salable product per unit of land area.** Any time your sales are going up while your costs are going down you have a win-win situation.

When you add in that the steer calf from a 1000-lb cow

can finish at 1100-lb at 24 months of age versus a 1600-lb steer from the 1500-lb cow at 36 months, you have a win, win, win scenario.

In Dr. Williams' analysis, the 1000-lb cow whose calf is taken all the way to a forage finish and sold at today's grassfed prices had a $229 profit advantage over the calf from the 1500-lb cow. This was a total increased return for the entire 400-acre ranch of $32,147 or just over $80 an acre in increased return.

*I generally recommend that novice grass finishers start with late spring finished heifers.* Quite frankly, very few graziers have the combination of grazing skills, quality pasture and genetics to finish a growthy North American steer by 24 months on an all-perennial system. Due to our large body phenotypes, most North American steers will require 28 months and a summer or winter annual to finish.

Not only are heifers much easier to fatten, most graziers feel they are less risky from a marketing standpoint. If you are just developing a market for your beef there are a lot more marketing options for a 1000-lb female than for a similar weight steer. If worse comes to worst, you can always sell her as a young stocker cow which, depending upon the cattle cycle, can be a pretty pricey animal.

The James Ranch in Durango, Colorado, is a pioneer grass-finisher and yet after a dozen years still only direct-markets grass-finished beef from heifers selected out of their replacement program.

---

**Most American graziers consider finishing weight cattle as just heavy stocker cattle. This is not the case. They are an entirely different class of livestock.**

**Anibal Pordomingo explained the big difference between the two is that the stocker cattle's weight gain is flesh and bone; whereas the majority of the finishing animal's weight gain is fat.**

"Grass finished animals are risky in that if you can't sell the meat there is often no other market for them," David James said. "As a result we prefer to sell our steers as feeders on the commodity market."

The James' 900 cows are on upland Forest Service range and the replacement and finishing heifers are grown out on irrigated, river bottom land.

Nutritionist Dick Diven said treating first calf heifers similar to finishing animals is a good program to follow for a high reproductive rate.

"The amount of energy needed for a heifer to produce milk, grow and breed back exceeds that found in the very best spring ryegrass pasture," he said.

"The heifer has to be able to mobilize her bodyfat reserves to get the energy she needs. This means she must have the fat cells in place and a diet to fill those reserves before she calves."

Mac Magruder agrees about the marketing risk in a new grassfed startup. He has a several thousand acre extensive, rangeland cow-calf operation combined with a 300-acre irrigated bottomland finishing operation in Mendocino County, California. Magruder is the primary beef supplier to the world famous Chez Panisse Restaurant in Berkeley, California.

Magruder has been working on grassfed beef for 20 years and like James started with culled replacement heifers. The year 2004 was the first year he was able to sell all of his 300 beeves as premium-priced, grass-finished animals.

"You've got to be willing to make a commitment to quality from day one," he said. "The commodity market will pay you 50 cents a pound for a 1200-lb grass-finished steer, so you better have the quality to sell it to a premium priced market before you start to finish steers."

You'll read more about the advantages and disadvantages in finishing heifers in the chapter on Start with Heifers.

Hopefully, you now have a basic understanding of the agronomic and genetic problems involved in grass finishing. This should better help you understand the history lesson from the earlier chapter.

## Let's Review

To emphasize the importance of this information, let's review some key points:

■ With grain feeding, the primary problem is the animal getting too fat. With grass, it is getting the animal fat. The animal genetics designed for one system will not be the best fit for the other.

■ For an animal to fatten requires the forage to be highly digestible and rich in soluble carbohydrates (sugar). Soluble carbohydrates are not only determined by the plant species but the stage of growth and the level of soil mineralization.

■ Finishing pastures are traditionally high in legumes, not only because of their higher digestibility, but because the soil nitrogen they create is in a slow release form that does not cause the plant proteins to spike as artificial nitrogen does.

■ Artificial nitrogen can frequently cause daily gains to fall in half.

■ Any weight gain in a fully grown animal must come primarily from the creation of fat.

■ 75% of the carbohydrates the animal eats will be consumed in body maintenance.

■ The slower the animal gains the more of it will go to maintenance. This is why the animal must gain in excess of 1.7 lbs per day to exceed this maintenance threshold.

■ To produce intra-muscular fat requires an average daily gain in excess of 1.7 lbs per day for at least 60 to 90 days prior to harvest.

■ To create intramuscular fat requires that we use a forage that is at least 65% digestible, at least 20% dry matter, with a protein level range that does not exceed 18% and a soluble carbohydrate content of at least 15%.

■ Due to the shortening of the day length, perennial grasses start to shift carbohydrates from their leaves to the roots to prepare the plant for over-wintering. This drops the carbohydrate percentage in the pasture too low for marbling.

■      At least 60 days of high gains are needed to finish an animal.

■      Fully grown animals once fattened tend to stay fat as long as they are fully fed.

■      With the exception of Eastern gamagrass, warm-season perennials have little place in a grass-finishing program.

■      With grass-finishing animals, heat is a bigger problem than cold.

■      A high quality beef eating experience starts when the animal is born.

■      It is critical that the animal have enough milk to be fully fed in the first 13 days of life.

■      Another "critical moment" in a beeve's life is when it is 65 to 70% of its mature weight (700 to 800 lbs). This weight is typically reached one year after calving. Beeves that are gaining weight at this time will form the intra-muscular fat cells that will be filled with fat during the subsequent finishing.

■      Endophyte-infected fescue should never be used in the final finishing phase due to the highly objectionable off-flavor it produces in the meat and gains are generally poor.

■      Because so much of the eating quality of the animal is determined prior to the animal reaching finishing weight, most gourmet-oriented, grass finishers are loathe to buy-in calves and yearlings from unknown producers.

■      Delay weaning until the dry matter has risen to the point where good gains are possible.

■      It is far better to bring the calf through the winter and difficult late summer period on the cow.

■      Never buy thin calves and do not try to finish any of your own home-raised calves that are significantly thinner than their herd mates.

■      The use of compensatory gain has no place in a gourmet program.

■      For the tenderest meat, the animal should never be

allowed to lose weight at any time in its life.

■　　　With a birth-to-finish program you will have three distinct classes on your ranch at the same time: cows with calves, stocker yearlings and finishing two year olds.

■　　　All grass finishing programs should start with weighing your cows. Without this critical number you will not have a target for your finished weight.

■　　　Weigh all your cattle every 90 days and all animals that are gaining less than 1.3 lbs be culled and sold at auction rather than taken on to grass finish.

# CHAPTER 5

# Horses for Courses — Using the Right Genetics

As noted previously, perhaps the most lasting damage massive grain feeding will have on the North American beef business will be what it has done to our cattle's genetic makeup. Very few yet realize just how bad our current genetic situation is.

The shift to late maturing, growthy cattle has been economically devastating to the cow-calf producer because it has destroyed fertility. If we restore fertility, we automatically restore the genetics that will finish well on grass. A concentration on reproductive efficiency serves the ends of grass finishers equally well.

The world renown South African animal scientist, Dr. Jan Bonsma, said in his excellent book on cattle breeding *Man Must Measure* that the most neglected question in livestock genetics was:

"What is the purpose pursued by man in breeding domestic animals?"

He said the only reply possible was: "His purpose is to produce animals which at the lowest possible cost and expenditure of labor give the highest possible and longest lasting returns."

He said that in the search for exceptional animal performance what has been most overlooked is that in the land-based grazing business, productivity is measured by the maximum amount of beef produced per unit of land area — not per unit of livestock. Bonsma said ***when this land-based criteria is used the most important factor in genetics is fertility.*** He defined fertility in females as having a regular estrus, normal ovulation, ready conception on mating or artificial breeding, a normal gestation period, normal parturition, giving birth to a normal calf and ultimately weaning a good calf.

Harlan Ritchie, professor of animal science at Michigan State University, said a Canadian study found that a 1% increase in the reproductive rate was worth twice as much as a 1% increase in price. Even more dramatic, an Australian study found that reproductive rate was 50% of the total net economic value produced in a beef cow-calf operation.

Ritchie said that by selecting for cows high in fertility you automatically selected for cows that were lower in maintenance. In other words, these two most valuable economic traits are always going to be in train and reinforce each other.

For example, cows that are high in reproductive rate are invariably moderate in milk production. ***You cannot have high fluid milk production and a high rate of reproduction.*** They are the opposite ends of the reproductive teeter-totter. Just as devastating as the drop in reproductive rate in high milk producing cows is the increase in body maintenance costs. Ritchie pointed out that cows that are high in milk production are also high in maintenance costs ***even when they are dry!***

The major problem in bovine feed efficiency is the high body maintenance cost of the brood cow. Feed efficiency is the ratio of feed input to salable product output. Twenty-one percent of the energy a chicken eats gets converted into edible protein. This compares to only six percent in beef cattle. And, the brood cow is largely why the beef ratio is so low.

Half of the total energy required to produce a pound of edible steak is consumed just for the maintenance of the cow.

In other words, half of the feed energy consumed in the cow, the bull that bred her, the calf, the yearling and the finishing steer is used just to keep the brood cow breathing air in and out. If you want to improve the profitability of your cow herd you have to start with cow maintenance costs.

Ritchie said that our attempts to increase beef cow efficiency by increasing milk production, weaning weight and yearling weight have all actually worked against us economically because they have all increased cow maintenance costs.

He said the type of cows we should be selecting for to have a low body maintenance cost would be:

- Lower in milk production.
- Low in visceral organ weight.
- Low in body lean mass.
- High in body fat mass.

Interestingly, he said these are the same traits you would be selecting for if you were selecting for highly fertile cows.

Another piece of serendipity for grassfed beef producers is that *animals that breed well on grass will also fatten easily, and therefore, finish well on grass.* This is why grass finishing and increased cow-calf profitability go together so well. They both benefit from the same kind of genetics.

Bonsma said after cows have proven their fertility by breeding regularly for several years they can start to be sorted for functional efficiency and as a possible base for future genetics. In other words, cows should prove themselves as fertile over time before they should be considered as possible seedstock genetic material.

Bonsma believed that trying to produce seedstock in soft environments with the heavy use of supplemental feed and hay so masked the environmental adaptation aspect of genetics as to make economic selection impossible. He believed that nature's genetic selection occurs fastest in challenging environments. Animals selected to do well in hard environments do exceptionally well in soft ones.

However, the reverse is not true.

He described functional efficiency as the combined effects of all factors through gametogenesis, libido, ability to copulate, estrus, ovulation, fertilization, embryo survivial, gestation, parturition and mothering ability of the cow.

He said low fertility was the number one criteria for culling cows for slaughter, and commercial cattlemen generally accepted this and cull their cows well.

While fertility has generally been thought to not be that heritable, Bonsma said a big reason for this is that the bulls used by commercial cattlemen have not been subject to the same strict culling criteria as their cows. In fact, many of the traits

---

### Using Ulta-Sounded Bulls

**On James Girt's ranch in Clatskanie, Oregon, only bulls that have been ultra-sounded for tenderness, marbling and rib-eye shape are used. Just by switching to a high marbling bull his meat has increased from High Select to Low Choice.**

**He wants a heavily muscled, full-bodied bull with short legs. All of his current cattle are black Angus. Like many, he has let some of his cows get too big and is scrambling to get back to a 1000-lb cow as fast as he can.**

**The cows calve in June. The calves are left on the cows through the winter and weaned directly onto green spring pasture in late March.**

**"Not weaning in the fall is a little harder on the cows but it is a whole lot easier on the calves," he said.**

**"The cows still have plenty of time to recover their body condition on the spring pasture before calving again in June."**

**The cows and calves are over-wintered on hay fed under cover due to the area's extremely rainy winters. His ranch gets 60 inches of rain a year but very little from July until September.**

**Grasses used include Alta fescue, orchard, Reed's canary and perennial ryegrass.**

seedstock producers select for are in direct opposition to the production of highly fertile bulls.

He said that when highly fertile cows are bred to highly fertile bulls, fertility is significantly inheritable. If no feed supplementation is employed, significant genetic progress can be made just through selection for fertility and culling.

*The heart of animal genetics is the genetic adaptation of the animal to the environment it will have to live and work in.* Bonsma said the primary problem with bulls is not complete sterility but sub-par fertility. This is most often caused by gonadel hypoplasia or underdevelopment of the testes, which is an inherited defect that can be passed on to both male and female progeny through a recessive gene.

In the hypoplastic bull the testes are small and of different sizes often with one testical lower than another. In a highly fertile bull, the testes are well developed and are perfectly equal in size and are parallel.

As a recessive gene, hypoplasia can be passed on to a bull's daughters and cause their sexual organs to never fully develop. Hypoplasia has been particulary devastating in the dairy industry because of the heavy use of artificial breeding from only a few bulls.

Bonsma said that one very popular Ayrshire bull in England nearly destroyed the whole breed when he passed his genetic defects to his many daughters. Two show-winning bulls were subsequently found to be the originators of the Swedish Mountain breed's reputation for poor fertility.

Here's where buying "performance tested" bulls can create real problems: Bonsma said *many bulls that win bull gain tests have hypoplasia as there is a strong correlation between high average daily gains on concentrate diets and low fertility.* In other words, using gain tests as a major selection criteria usually selects for sub-fertile bulls.

There also is a strong correlation between large frame size and low fertility. Steered calves grow longer legs than those that aren't. Bulls with long legs are usually sub-fertile.

Interestingly, Bonsma also found a strong correlation between hair color and hypoplasia. He said the more strongly pigmented animals have better developed genital gonads.

Red cattle should be a deep red. In red-brown cattle, the cows should have an even body color, but fertile bulls will show a darkening of the hair on the head, neck, crest and lower part of the body. In white haired cattle, the fertile bulls have a creamy, yellowish color and the infertile bulls are more nearly pure chalk white.

Estrogen inhibits hair growth. Fertile females have fine hair. Testosterone causes hair to be thicker, coarser and darker.

Testosterone causes the skin to be thicker, especially in the region of the neck and shoulders and the hair is particularly dark in this region. These male hair markers will not develop in the absence of testosterone and are a good marker for a well-functioning hormonal system.

Another genetic defect in bulls is prepuse prolapse. Bulls that prolapse develop phimosis which tightens the foreskin so that it cannot be drawn back from over the glans. This is a

---

**Low Yielding Cattle Are Costly**

**Washington state grazier George Vojkovich said he made a mistake by not paying attention to the phenotype of the beef animals he was buying. He had been using a mixed herd of Red and Black Angus cows. Now he has 150 mother cows. While he likes the color of the Red Angus, he hasn't liked their lower meat yield due to their lighter muscled back ends.**

**"After hearing Gearld Fry talk, I started keeping individual meat cut-out records on my cattle, and just as Gearld had said, there was a huge yield difference from one animal to another, even though they weighed the same. When you are selling meat for $3.00 a pound, a hundred pound difference in meat cut-out is serious money."**

particular problem in polled breeds and in Bos indicus breeds such as the Brahman. Bulls susceptible to this problem tend to have pendulousness of the sheath as well as the sheath opening. This is a major problem in rangeland environments.

This and hypoplastic testicals are problems that can be easily identified through visual examination and can just as easily be eliminated through selection. Unfortunately, many seedstock producers are loathe to cull their expensive animals and prefer to pass them on to unsophisticated commercial bull buyers.

It is through this common practice that most of the fertility problems are entering the commercial cattle industry. In other words, the seedstock industry is the source of most of the commercial cow-calf producer's problems.

Endocrine imbalance also has a marked influence on the reproductive ability of cows and can be caused by overfeeding

---

**Start with the Right Genetics**

**Mac Magruder's herd in Potter Valley, California, was originally Shorthorn, but he couldn't sell everything as grassfat 20 years ago. He crossed them with Angus due to market prejudice against the fast-fattening Shorthorns.**

**"I made the same mistake a lot of people did and used bulls that were too big and got my cows too big. Today, I wish I had my original herd back."**

**Magruder said grass finishing is much easier if you start with the right genetics, which are mid-sized, easy fattening cattle. He is downsizing his herd's frame size as fast as possible.**

**"Large cows just do not do well in the infertile hills of Northern California where low soil selenium levels are a particular problem. The type of cattle that will work well on our hills are the same type of cattle that will grass finish."**

grain. However, Bonsma believed that some females have a hereditary predisposition to suffer from an endocrine imbalance at slight environmental stress. This is why seedstock animals should be selected from hard environments as they will quickly weed out cows susceptible to this problem.

Grain feeding is commonplace on the majority of North American seedstock farms. One of New Zealand's leading bull studs documented the effect of grain supplementation on bulls and subsequent semen quality. They found that bulls that had been developed entirely on grass with no supplemental grain had a live semen count of 80 to 90% with 85% being average. These bulls also only had two to three percent abnormal sperm. An abnormal sperm is a genetically defective sperm cell.

Bulls that had been supplemented with unprocessed grain had a live semen count of only 65 to 70% with 15 to 18% abnormal sperm cells.

Bulls that had been supplemented with heat-treated commercial feeds had a live semen count of only 60 to 65% and the abnormal semen count was above 18%.

Now, consider that 10% of the semen die in the freezing process and you can see that the semen from grain supplemented bulls is borderline for settling a cow.

According to Bonsma, when cows are fed grain, biological and hormonal processes are accelerated. Follicular activity increases in the female and more inseminations are required for conception.

Cows fed grain and maintained in a high body condition require far more veterinary attention than those maintained on grass alone. Their fertile lifespan is reduced as they succumb to laminitis, arthritis and uterine infections.

Fattening bulls creates fat in the neck of the scrotum that insulates the testes. The heat-exchange system in the spermatic cord becomes ineffective with the presence of fatty tissue, resulting in a reduction in sperm mobility and an increase in abnormal spermatozoa.

"Bull melt" is a common problem for commercial beef

producers who buy over-fattened, grainfed, seedstock bulls. Not only do these animals immediately lose condition but normally take six months to a year to readjust their rumens to a grass-only diet. Some never readjust.

Keep in mind, a primary reason to castrate and spay animals is so they will fatten easier. Castrates also grow more frame and have long smooth, feminine muscles. A long-legged bull that is blocky and has poorly defined muscling is always a low fertility animal but often wins livestock shows.

Bonsma said the highly fertile cow is of slender build. Her neck and front portion are slim and her coat is sleek. She will not be fat or have masculine defined muscling. Her shoulder blades are lean and loose. The cartilaginous ends of the shoulder blades can be seen to move freely above the highest points of the thoracic vertebrae. A highly fertile cow has a big butt. A

---

**Low Stress Cattle Handling**

**There had been a belief in Argentina, France, and some other countries that docility was an inherited trait primarily due to the breeds used. However, a joint Argentine/French research program into animal behavior at the INTA Food Technology Institute in Buenos Aires found that this was not the case.**

**Using Argentine Angus, Argentine Zebu crosses and full blood French beef cattle, the researchers found that all three breed groups were extremely tame and gentle when handled frequently and correctly by humans. They also found that all three breed groups were equally wild when raised in the absence of human handling.**

**While this finding was not what the French seedstock producers wanted to hear, the researchers concluded that tameness, docility and low pH tender meat are all man-made attributes and have little to do with breed genetics.**

highly fertile bull has big shoulders. They are mirror images of each other. Neither is blocky like a steer.

Let's quickly review the best ways to produce sub-fertile animals.

■ Select animals in a soft, unstressful environment.

■ Select animals with abnormally high rates of gain.

■ Select animals that have a large frame, long legs and a blocky body.

■ Feed them supplemental grain to get and keep them fat.

Unfortunately, this is the production protocol at the majority of North America's seedstock ranches.

Seedstock breeders are not stupid. They do what they do because that is what they are paid to do by their customers. Fat cows and bulls are seen as signs of good, not bad, stock-man-ship. As long as fat bulls top the sale, they are going to sell grain-fattened, subfertile bulls.

Gearld Fry is a long-time student of Bonsma's work. He has worked in the seedstock industry as a reproductive physi-ologist for most of his life and said *every year the cattle he tests have gotten phenotypically worse, more over-condi-tioned and less fertile.* He said that most North American seedstock cattle were now so bad that the only hope he sees is for commercial cattlemen to produce their own bulls. He said this was particularly true of cattlemen who wanted to produce high quality, grassfed beef and direct market it to the consumer for a premium price.

Fry works as a consultant to many of America's leading grass-finishers. He teaches his clients to identify animals in their herd that have superior meat eating characteristics from highly apparent visual cues. For example, selecting for fertility naturally increases the rump size in both males and females.

An increase of one inch to the length of the rump adds 50 lbs to the carcass weight. An increase on one inch to the width of the rump adds an additional 50 lbs to the carcass weight. The two of these together are 100 lbs more high value meat. At today's direct market prices, this selection alone can add as much as $300 to the value of the animal.

Here are some other interesting correlations he found:

■      Beef flavor is highly correlated with glandular function. Animals that have a dark, oily streak down the middle of their back will have superior flavor.

■      Animals that have a shiny, oily hair coat will also have superior flavor. Animals with dull, rough haircoats are invariably a poor eating experience with an objectional off-flavor and should be sorted off and sold as feeder cattle rather than finished.

■      Meat tenderness is also highly correlated with milk butterfat. The higher the butterfat in the cow's milk the more tender the meat is in her calf. This is why the Jersey and the French cheese breeds have such tender beef.

■      Meat tenderness is also correlated to bone fineness. Cattle with hard fine bones have exceptionally tender meat.

■      Select for animals with flat rather than rounded rib bones for exceptionally tender meat. A flat jawbone is another marker Fry has found for genetically tender meat.

■      Long legs with no meat on them mean a low cutout carcass. To have a high meat-to-bone ratio the animal should always be thicker in body than the length of its legs and the meat on the animal's rear leg should extend down to where the leg bends.

Fry said an animal's potential for eating quality can be largely ascertained by the time it is six to eight months of age. This makes knowing the visible clues related to meat quality all the more valuable for those who buy rather than breed cattle for grass finishing. Gearld Fry is not a fan of crossbreeding. Surprisingly neither was Jan Bonsma. I say surprisingly because Bonsma developed a composite cattle breed — the Bonsmara.

Bonsma believed that breeds were only justified as environmental adaptations to specific geographic niches. The black Angus for example was selected for dim, foggy valleys in Scotland and not the hot sunny plains of Kansas.

While he thought the idea of a single breed fitting in

everywhere on a continent to be ridiculous, he found little justification for there needing to be more than one breed developed for any particular climatic niche.

He said the only reason he developed the Bonsmara breed was because the heat-adapted, tropical Bos indicus breed had never been selected for the Western economic traits of efficiency of food utilization, fertility, milkability, growth and quality eating.

Similarly, all Bos taurus breeds had been developed in the high latitudes of Europe and had little resistence to the tick-carried diseases found in the southern United States and many of the countries of the Southern Hemisphere.

He found that the European Bos taurus breeds could be selected for heat tolerance but could not be selected for resistance to tick-borne diseases. Hence the Bonsmara, which was 5/8ths Afrikaner and 3/8ths British breeding.

Again, this breed was developed to provide natural resistance to ticks not to heat. You can select for heat-resistant animals from within the Bos taurus breeds. The Senepol is a good example. The Ruby Red Devon is another.

Bonsma said a big problem cattle producers in the New World have had is that we brought breeds of cattle that had been bred to be productive in the high, cool latitudes of England and Scotland and expected them to work just as well in the lower, warmer latitudes of North America.

British cattle are most comfortable between 40°F and

---

**Early Lessons**
**Tom Gamble of St. Helena, California, learned early on the importance of genetic selection in grass finishing.**

**"The first steer I bought to finish was what I thought was a black Angus steer. However, it was actually an Angus-Chianina cross. That animal just grew and grew and grew."**

65°F. When the temperature goes above 70°F, the British breeds start to suffer from hyperthermia (heat stress).

Following Bonsma's observations and recommendations, here are some guidelines for what to genetically look for in cattle if you live in an area of the country where daytime summer temperatures are far in excess of 65°F.

## Phenotype

Animals that are adapted to extreme heat conditions have a large surface area per unit of weight. The smaller the animal the larger the surface area per unit of weight.

*Cattle that have to live in hot, humid conditions must be smaller than those of the same breed in more temperate or semi-arid regions.* Also, because digestibility of the grass goes down as temperature goes up, cattle in hot and humid areas have to have much larger rumens than in temperate grass areas. This means they cannot have the smooth underline of a show cow. Instead, think of the large distended waist of a Jersey cow.

Nature tries to keep the forage game between the North and the South relatively even by off-setting natural advantages with equal and opposite disadvantages. For example, the warm climate that allows a green forage year around in much of the South has an offset in the forage being less nutritionally dense.

Nutritionist Dick Diven described this phenomenon in terms of a cake and a can of chocolate icing. He said both the North and the South have about the same amount of icing for their forage cakes.

In the North the cake is smaller due to the shorter season but its icing is really deep and rich. In contrast, the farther South you go the cake gets bigger due to the longer growing season but the fixed amount of icing has to be spread much thinner to cover this larger cake.

Diven said an example of this can be seen in irrigated alfalfa hay production. In Idaho ranchers can cut ten tons of alfalfa in four cuttings, but it requires eight cuttings to get the same ten tons from the longer growing season of Arizona.

This is also why it has been noted that Southern stocker cattle can go North and do better but Northern stocker cattle can't go South and do as well.

A genetic result of this less nutritionally dense forage is that as animals live closer to the equator they will become smaller in phenotype as they adjust to the falling nutritional density of the forage. This is why a trophy deer in Texas weighs 150 lbs and one in Canada weighs 300 lbs.

Diven said this is a good reason to ***never buy in brood stock from outside your own ranch's latitude.*** Cattle purchased dramatically farther North or South from your ranch will be genetically mis-matched in phenotype for your area's nutritional density.

## Skin Thickness

Cattle primarily cool themselves by increasing the blood flow to the surface of their bodies. This is called vascularization. Cooling through the skin is far more important than panting in cooling the animal. In fact, the effort of panting actually increases the heat production of the animal.

Animals with thick hides which allow for more blood flow are far more heat tolerant than animals with thin hides. The Red Devon has the thickest hide of all the English breeds. This is why it was so popular in the Deep South region.

Thick loose hides also help an animal repel flies and ticks.

### Skin Thickness by Breed

| | |
|---|---|
| Red Devon | 8.15 mm |
| Hereford | 6.77 mm |
| Friesian | 6.08 mm |
| Zebu | 5.77 mm |
| Angus | 5.75 mm |

## Hair Sheen and Color

Animals with a short hair coat and a glossy sheen reflect far more light (and heat) than animals with long hair and no sheen. For example, an animal with a short, red haircoat reflects about 12% of the sunlight falling on it versus only four percent for a dull red, long-haired haircoat.

The following shows the percentage of light reflected by glossy haired animals versus non-glossy animals.

| Color | Percent of Light Reflected |
|---|---|
| **White** | **15%** |
| **Yellow** | **14.5%** |
| **Golden Yellow** | **14%** |
| **Light Red Brown** | **13.5%** |
| **Red Brown** | **13%** |
| **Dark Red Brown** | **11%** |
| *Non-glossy Breeds* | |
| **Red-Brown Shorthorn** | **4.6%** |
| **Black Angus** | **3.7%** |

The amount of difference between a dark red haircoat and a light red haircoat is considered statistically insignificant.

Black haired and black hided animals such as the Angus absolutely have to have access to shade in hot weather. These cattle were developed in cloudy, misty Scotland and do not do well in open, sunny pastures.

At 90°F, Bonsma said the hide temperature of a black-haired, black-hided animal will be between 113° and 122°F.

Bonsma estimated that the amount of solar energy falling on a 990-lb Angus cow in a six-hour day is enough calories to boil 13,162 gallons of water. Is it any wonder they want to stand in the pond all day? However, he said animals that stand in water up to their belly or in the shade all day, do not eat and soon become chronically under-nourished. This not only sets them up for a poor reproductive rate but makes them suscep-tible to flies, parasites and disease.

Cattle that develop a high internal body temperature from either fever or hyperthermia will likely suffer permanent pituitary damage. These animals can be recognized by their rough summer haircoat with no sheen. The pituitary is what makes cows cycle.

No pituitary. No calves.

With British breeds in hot summer regions, *it is very important that the animal be open during the worst of the summer's heat to prevent abortion* and what Bonsma calls "tropical degeneration" or small calves.

"Miniature calves can be produced when cows are mated early in spring and are pregnant through the summer," Bonsma said.

This tends to be more of a problem with male calves than females because male calves have higher — and therefore hotter — metabolisms.

A June 21 target calving date eliminates this problem and also moves the breeding season out of the summer's heat to late September.

## Hide Color

While black hair is definitely a no-no in open, hot areas, black skin is a plus as it protects the animal from harmful ultra-violet rays. Animals at elevations above 5000 feet must have a dark hide to prevent skin cancers. Good high elevation breeds are Angus, Simmental and Brown Swiss.

In hot, humid areas, Bonsma said you want an animal with light colored hair but a dark hide. A excellent example is the Jersey.

I read research from Jamaica that found the Jersey was 99% as heat tolerant as the Zebu in their hot, humid climate.

## Hair Shedding

North America's large land mass not only produces hot summers but cold winters. This means that in the temperate areas the animal must hair well for winter but then must quickly shed its winter coat in the spring to avoid heat stress.

"There is no single factor that gives such positive results as the selection for early hair-shedding in spring," Bonsma said. Air is an extremely poor conductor of heat. Long hair traps air close to the body and makes it impossible for the animal to cool itself.

British cattle that don't quickly shed their hair coat can suffer from hyperthermia and will frequently abort their calves. Calves that don't abort are born permanently damaged and are very small and extremely poor doers their whole lives.

Calves do not develop an ability to deal with heat until

---

### Things to Look for in Your Herd

- Calves should have a navel cord string at birth.
- Heifers should be cycling by 10 months of age.
- Heifers should conceive at 14 months of age.
- A bull or cow that refuses to shed long hair should be culled.
- Bulls and cows should have pigmented skin and black hooves.
- Short, thick hair is more fly and tick repellent.
- Dull, long hair attracts flies and ticks.
- A highly fertile bull will have darker hair on the lower half of his body.
- A bull and cow should have a wide mouth and large nostrils.
- The cow and bull must have a large gut capacity.
- Height destroys reproduction. A tall, long-legged animal will be slow to mature and difficult to finish.
- Cattle with long body lengths are normally high maintenance animals.
- The shorter the neck of the bull the higher the male hormones.
- As the bulls' neck gets shorter his progeny's rump tends to get wider.

**Gearld Fry**

they are a year old and will suffer terribly through their first summer with low gains and possible lasting physiological damage if they have long hair. Most stocker graziers know that buying long-haired, fresh-weaned stocker calves in the summer is a sure recipe for a high death loss and little to no gains on those that do live.

*Animals that do not shed their haircoat rapidly are almost always sub-fertile animals.* A high estrogenic status and regular cycling are prerequisites to early hair shedding and a clean, glossy summer coat. Therefore, using this one criteria for culling removes most of the problems in a cowherd.

"Early hair-shedders are well adapted to their environmental conditions, they reproduce regularly and have high production records," Bonsma said. Culling on this one criteria of fast hair shedding can remove a host of problems from your herd in hot summer regions.

When you calve is much more critical in the South than in the North due to the influence of day length.

Beef cow nutritionist, Dick Diven, said that as you near the equator the variance in the day length from summer to winter is diminished. The subtropic zone has longer days in winter and shorter days in summer than the northern temperate zone. This lessens the natural reproductive stimulus of the change in season.

Bonsma, observed that *sexual activity in cattle peaks at the equinoxes* (September 22 and March 21 in the upper latitudes). This is one reason why calving close to June 21 (the longest day of the year in the Northern Hemisphere) has been found to increase calving percentage so dramatically. (15% to

---

**Selecting for birth weight, milk, height, length, weaning weight or any other one trait is single trait selection. Selecting for reproduction and low maintenance creates all good needed traits.**

**Gearld Fry**

20% higher than winter calving.)

In the 1970s and 80s, crossbreeding became the thing because it was found that the crossing of two dissimilar genetics resulted in a genetic growth "kick" called hybrid vigor that increased average daily gain.

This was a cheap competitive advantage in a commodity environment where we "sell them by the pound" and aren't paid for quality, and, as long as we didn't keep the crossbred females for replacements. Unfortunately, we did.

Unplanned crossbreeding tends to quickly unwind the

---

**Concern for Lack of Quality Control**

**Organic meat marketing consultant, Mark Keller, said his biggest concern about grassfed beef was the lack of quality control over its eating quality.**

**He said not only was there huge variation in the quality of product between ranches but from one animal to another from the same ranch.**

**He feared the current grassfed boom could fizzle out due to poor eating experiences.**

**"Anyone can sell a product once, but repeat customers are what give you a business."**

**He said the only way he knew to take this variability out of the product was through tightly controlled genetics.**

**Keller said currently profits per head in grassfed beef ranged from $300 to $900 an animal.**

**While liveweight at harvest has a lot to do with this, a big part of this variation is due to meat cutout yield. "A three tenths of a percent increase in meat cutout yield increases the price received per head by $356.33 at today's grassfed retail price," he said.**

**"All that Gearld Fry has been telling you about the huge dollar return to genetic improvement is very true. The best solution is to tightly control genetics."**

very traits that identified an animal as a breed apart from others. The problem with non breeds is not that there aren't some good animals among them but that there is no predictability in what qualities its progeny will have.

This problem has gotten much worse because many so-called "purebred" cattle aren't purebred. The not-so-secret, secret of seedstock genetics is that many English breed breeders have secretly crossed their cattle with the Continental breeds to make them taller and faster growing. This has not only made them less fertile but has also destroyed their hard-won reputation for quality beef.

The occasional tough-eating animal is explained by the comment that there is now more variation within breeds than between breeds. If this is so, what purpose does having distinct breeds serve? *A breed is supposed to stand for a reliable outcome not an indefinite one.* What has been forgotten by nearly everyone in genetics is that people have to eat this stuff.

In the early 1980s, Fry met Jan Bonsma, and began studying his observations on the co-relation between body phenotype and reproductive problems. Dealing with many animals with reproductive problems every day allowed Fry to see first-hand what Bonsma was talking about and to train his eye to spot problem animals.

After 15 years of study, he said he can now walk through a rancher's cowherd and tell him past problems he has had with particular cows. He can do this due to the phenotypic aberrations he sees in the animals. He said the primary problem

---

**Start to Finish**
**Length of the growing-finishing period (which in a grassfed program starts when the calf is purchased or weaned until it is sold as a finished steer or heifer) should be kept as short as possible. This requires animals of suitable and similar genetics and frame.**

most ranchers have is that they have never developed such a "stockman's eye."

Agreeing with Harlan Ritchie and Bonsma, Fry said the *only two traits that make any sense to select for are repro-duction and body maintenance.* As Ritchie previously pointed out, if you select solely for reproduction you automatically get an "easy-keeper" cow and one that is easy to finish (marble). However, Fry has found that selecting for reproductive fertility also increases carcass meat yield in the high value cut areas.

Femininity in a cow is primarily expressed by the width of her rump. A wide-rumped cow is also an indicator of early maturity and easy fleshing, which are exactly what we are looking for in grass-finished animals.

Wide rumps are also important in the bull as well. As in the cow this is an indicator of early maturity and ease of fleshing. Fry said to keep in mind that 88% of the high value steak cuts

---

**Jo Robinson's Taste Tests - USA**

The most common complaint of judges in taste tests of grass-finished meat is "The juice and flavor didn't last through the chew." By the time the meat was ready to be swallowed, all the tasty fat and juice had gone down the gullet.

This was caused by 1) having too little lubricating fat to begin with; 2) being so tough that all the fat and juice were gone before the meat could be swallowed; or 3) a combination of the two, which was most often the case. Most of the steaks should be fatter and more tender for the typical customer to rave.

"There will always be those who are so in love with the idea of grassfed beef that they'll put up with meat that is on the dry and tough side," Robinson said. "But most people eat beef to enjoy it. Grassfed meat can be utterly delectable. We need to help producers attain that goal."

are in the rump. The wider the bull's rump, the larger will be its rib eye and the loin eye.

Interestingly, a large-rumped fertile female will produce a very masculine wide-shouldered son. Conversely, a large-rumped bull will produce a large-rumped highly feminine daughter. This is the win-win scenario of selecting solely on reproduction. You get the best traits in both the male and the female.

Selecting for bigger bulls is always a selection for delayed maturity. These bulls' progeny will grow frame rather than finish. Fry said *selecting solely for reproduction will naturally produce an animal that will mature early, fatten easily and have a high value carcass yield.* Fry believes that direct marketers should not crossbreed or buy in bulls from outside their own herd. This is because a direct marketer gets paid — and paid well — for a quality meat product. The typical seedstock breeder is not selecting for meat quality.

In replicating Bonsma's work, Fry has worked with a very specific series of ratios between height and depth, width and length to develop a computer program that can pretty closely predict the cutout meat yield of an animal at eight months of age. Also, Fry is consulting with several hundred grass-based breeders who are interested in developing a seedstock line of bulls just for grass finishers.

Needless to say, getting the right genetics will become easier in the future. However, be aware that most cattle out there today will not work for a gourmet grass-finished program, but some could get close to it.

# CHAPTER 6

# Myths and Truths About Grass-Finished Beef

During the December 2003 outbreak of BSE in Washington State, I heard a USDA spokesman say the USA didn't need to worry about losing the Japanese market because we were the only major source for grainfed beef.

"Grainfed beef has better flavor and is more tender than grassfed," he said with no qualification.

Of course, the Japanese apparently didn't hear him and quickly swapped their beef purchases to Australia.

This canard about the superior eating quality of grainfed beef has been repeated so long and so often that is has become scientific "fact" but there actually is very little research to back these "facts" up.

In the early 1990s, New Zealand decided to actually put these American promulgated "facts" about the flavor and tenderness superiority of grainfed beef to scientific analysis in a two-year study.

Dr. David McCall of Whatawhata Research Centre in Hamilton said some of the widely promoted and generally believed American "facts" about grassfed beef were:

1. Grassfed animals have lower carcass yields and weights.
2. Grassfed animals do not have marbling.
3. Grassfed animals have yellow fat.
4. Grassfed animals are less tender.
5. Grassfed meats have a different flavor and aroma.

**6. Grassfed meats have a different color.**

**7. Grassfed meats are highly seasonal in supply.**

In a study of American scientific literature on comparing the two feeding methods, the New Zealanders found that in no American study were animals purposely taken to the same degree of physical maturity before an analysis of the meat was made. They said this violates the basic principles of scientific research. If you are going to compare, you must compare apples to apples not apples to oranges.

*The degree of maturity is related to age and weight and the intersection of the two factors.* Because grainfed animals gain faster than grassfed animals, they reach mature weights earlier.

If a grassfed steer and a grainfed steer are killed at the same age or when the grainfed animal has become "finished" the grassfed animal will still be growing and will be unfinished.
In other words, if we use age it will be an apple to orange comparison.

The New Zealanders found that when cattle of similar genetic background are harvested at similar physical maturities the tenderness and marbling of grassfed and grainfed meat are virtually the same.

---

**Have a Cow, No Really!**

**At supermarkets and small farms across the country, herds of buyers are snapping up boutique meats: cows raised on grass, lambs allowed to frolic, even pork with a pedigree.**

**Though Americans are eating less red meat these days — about 4% less per person in the last two years — sales of "natural" meat have jumped 30% or more this year, producers say, due to everything from health worries to a raft of unsettling reports about the safety of commercial meat.**

**Wall Street Journal**

Now let's go back through the previous list with animals harvested at the same stage of physical maturity.

## 1. Carcass Yield and Weight

The New Zealanders said carcass weight is solely a function of growth rate and age. Carcass yield potential at maturity is determined mostly by genetics and secondly by feedstuffs. Animals with superior muscling will have superior lean meat yield regardless of what they are fed.

Big-boned, long-legged, narrow-chested animals have a lower meat-to-bone ratio than thick, full-bodied animals.

In total meat yield, fat cover is also important as fat is counted in total hot weight yield. The amount of fat present is dictated by sex and how heavy an animal is relative to its ultimate mature size.

A rule of thumb Gearld Fry uses is that an English breed, castrated male will typically "finish" at a liveweight that is approximately 100 lbs heavier than its mother. In other words, an 1100 lb Angus cow's son will "finish" at about 1200 lbs in weight.

If you harvest such an animal at 1000 lbs it will not have the

---

### Stone Age Diet

For more than two and a half million years, our prehistoric ancestors lived on a red-meat-lovers diet rich in lean, wild game. Obesity, heart disease and stroke were virtually unknown. Believe it our not, this stone age diet might be for you if you're watching cholesterol and don't indulge in sizzling steaks.

The difference is that stone age meats were full of heart-friendly Omega-3 and mono-unsaturated fats. In contrast, most beef today is grain-fed and high in Omega-6 fatty acids and saturated fats. There are ways around this. Look for grass-fed beef.

Dr. Emily Senay, "Healthwatch" CBS Radio Network

same yield percentage because it will not have the same degree of fat as a fully mature animal. The difference can be as dramatic an increase from 50% yield to 60% + yield.

Carcass weight and yield are important because the abattoir cost for the two animals is virtually the same. Therefore, the heavier the animal the lower its harvesting cost.

However, the New Zealanders warned that using European breed genetics can result in an animal whose mature "finish" weight is in excess of today's popular slaughter weights and are unlikely to finish at less than three years of age on pasture.

## 2. Marbling

Intramuscular fat tends to be one of the later fat deposits laid down. High marbling levels became apparent as the animal reaches its mature body size.

Marbling is also influenced by breed and genotype within the breed.

However, the primary determiner of marbling is how close the animal is to its mature size when it is harvested.

In the New Zealand study no difference was found in the

---

**Grass-finished Beef to the Rescue**
"As a society we are clearly in a state of nutritional crisis and in need of radical remedies. The statistics are sobering. After 30 years of seemingly solid advice aimed at lowering dietary fat, Americans have grown collectively fatter than ever.

"Today more than 60% of adults in the USA are classified as overweight or obese. So many children have become so heavy that pediatricians are now facing an epidemic of Type 2 diabetes and hypertension — diseases that are closely associated with being overweight and that were unheard of just a generation ago."
Time Magazine, September 2, 2002

---

degree of marbling between grassfed and grainfed animals harvested at the same weights. An Australian study found the same thing.

The New Zealanders said this is only common biological sense. Animals with the same carcass weight and fatness will have similar degrees of marbling. There is no magic in corn. Meat juiciness is directly related to the degree of marbling. Consumers have been found to mentally correlate juiciness with tenderness but they are not physically correlated in any way that can be objectively measured.

Genetically tender cattle are tender regardless of the degree of (or absence of) marbling.

## 3. Fat Color

In many countries, yellow fat is perceived to be from older animals and so is not preferred. However, this yellowness is actually B carotene, which can be metabolized to Vitamin A and is a major anti-oxidant. It could be that with sophisticated consumers, yellow fat will actually be preferred as it is in some continental European countries.

The New Zealand researchers found that yellow fat in grassfed animals was highly seasonal in nature and tended to be the highest in the spring and the lowest in summer. This seasonality was thought to be related to the presence of certain legumes. (Strawberry clover was found to be a prime producer of yellow fat.) In the study, animals harvested in mid-summer had fat color equal in whiteness to grainfed animals.

## 4. Tenderness

The New Zealanders found that if an animal is harvested prior to 30 months in age, nutrition has very little to do with meat tenderness. Tender meat is dictated primarily by the animal's genetics and the amount of collagen-connective tissue the animal has.

Through genetic screening, the New Zealanders were able to lower the number of tough meat eating experiences from one in four (the current USA average) to one in seven. While they said this is still way too high, it does show that meat tenderness is primarily a genetics problem and not a nutritional one.

While low meat pH has been correlated with meat tenderness, emotionally stressed animals with high pH meat have also been found to have tender meat but poor aroma and keeping qualities. As an animal ages the texture of the meat changes as the size of the muscle fibers increase, and some consumers may be put off the different mouth feel of this older beef. This was not found to be a problem with animals harvested at less than 36 months of age.

---

**Pasture-Fed Beef Tops Canadian Tenderness Test**
In a test of the tenderness of cooked beef measured physically and by a sensory panel for cattle finished entirely on pasture rated higher than those supplemented with grain.

The research project by the Crop and Livestock Centre in Charlottetown, PEI, Canada was primarily designed to see if pastured steers would have more CLA if supplemented with whole roasted soybeans rather than barley grain or TMR.

Steers fed pasture alone had the highest level of CLA and the highest levels of unsaturated versus saturated fat. The CLA level for the c9, Tll strain of CLA that has been implicated in the reduction of heart disease and cancer were nearly 60% above the soybean supplemented steers and 44% above the TMR fed steers.

The Research Centre concluded that pasture finished cattle can obtain acceptable carcass weights and finish and yield beef that is more tender than conventionally finished beef and contains almost twice the concentration of CLA.

---

## 5. Flavor and Aroma

Beef flavor is primarily attributable to compounds (fatty acids) found in beef fat. As an animal increases in age and weight the fat content increases and so does the flavor.

While the diet an animal is grazing can cause changes in the fatty acid composition, these changes are very subtle and may not be detected by most adult consumers.

In the New Zealand study, animals that had been finished on ryegrass to the same degree of fatness as grainfed animals were difficult for most consumers to distinguish between for aroma and flavor.

A taste test at Auburn University found the same thing with ryegrass cattle. However, there consumers objected to the meat taste of cattle finished on fungus-infected fescue.

## 6. Meat Color

The color of meat is related to the level of pigmentation (myoglobin) present in the muscle.

When beef is cut the myoglobin oxidizes, giving rise to a bright cherry red color and a process known as "blooming."

As animals reach mature weight, the redness increases.

---

**Grassfed Meat Has a Different Aroma**

**Dr. Gabriela Grigioni of Argentina's Food Technology Institute said that grassfed beef has a different aroma than grainfed beef.**

**She said that grassfed beef is rich in anti-oxidants that protect the very unstable unsaturated fats from deteriorating and causing a rancid smell. Therefore, most consumers find grassfed beef to have a "fresher" smell that they like.**

**She said it was important that grassfed beef producers never use chemical weed control on their pastures as this chemical smell will often come through in the meat's aroma.**

---

High grain rations do indeed produce redder meat than grassfed animals as grain feeding alters the natural level of pigments.

This color shift was found to occur after the animal had been on grain for six weeks.

In the New Zealand study, this slightly less red meat color and the (seasonal) color of the fat were the only two differences found between grain-finished and grass-finished beef harvested at the same mature weights.

The New Zealanders said the greatest variation in the eating quality of all meat lies in the cooking skills of the consumer. Learning to cook all meats at lower temperatures for longer periods would greatly add to their eating quality regardless of the feedstuffs used.

### 7. Seasonality of Production

Here is where the feedlot system really has a clear advantage, but as the New Zealanders point out, only at a significantly higher cost.

Grassfed beef can also be produced on a year around basis through the use of supplementary annual forage crops. This is known as a forage chain and will be discussed in a later chapter.

### The Amazing Benefits of CLA

Conjugated Linoleic Acid (CLA) is a recently discovered essential fatty acid with some pretty amazing side properties. Animal studies suggest that it is:

**1. Anti-carcinogenic**
**2. Helps prevent obesity**
**3. Anti-diabetic**
**4. Anti-antherosclerosis (heart disease)**

Dr. Tilak Dhiman of Utah State University is the foremost researcher into Conjugated Linoleic Acid (CLA) in the USA and a noisy champion of grassfed beef. He said that it is a shortage of production more than a lack of market that is holding the industry back.

He noted that American organic food sales have grown from $7.5 billion in 2001 to $18.4 billion in 2005 and are projected to reach $23 billion in 2007.

"Currently, the organic food section is the most profitable section in today's supermarket. Even Wal-Mart is bringing out an organic food line," he said.

"And yet, this is happening without any conclusive research that organic food is healthier for you. *We have a far stronger and better researched story in grassfed products.*"

Some of grassfed's benefits in comparison with the grainfed product are:

**500% more CLA**
**400% more Vitamin A**
**300% more Vitamin E**
**75% more Omega-3**
**78% more Beta-carotene**

In animal studies (humans are not used in cancer studies for obvious ethical reasons), 11 out of 11 had found CLA decreases cancer.

Four out of five have found a decrease in body fat.

Two out of two have found a decrease in heart disease.

Six out of six have found increased immunity to disease.

Two out of two have found increased bone density.

And three out of three have found a decrease in adult diabetes.

A French study of 360 women found that the higher the CLA level was in their breast tissue the lower their incidence of breast cancer.

Research has shown that CLA not only reduces the incidence of cancer in animals but actually suppresses the growth of cancer cells.

"I am convinced that grassfed foods are not only preventative but regenerative as well," he said.

While artificial CLA is now available in a pill form, natural CLA from animal products has been found to be 600% more effective in fighting cancer.

He said that CLA was an additive. In other words, eating grassfed beef and grassfed cheese and milk all helped accumulate CLA in body tissues.

"This is a much more exciting story than organic food products have to tell. Therefore, I believe grassfed products will have a much more explosive growth curve."

He said he was not denigrating organics as personally he would like his meat and milk to be both organic and grassfed.

"I believe an organic 100% grassfed product would be the ultimate in healthy eating."

He said these health findings should greatly benefit the growth of grassfed beef.

"CLA in a ruminant product increases linearly with the increase in fresh pasture consumption," he said.

"In milk, it takes 25 days on pasture to get to the maximum level but only five days for this level to collapse when the animal is removed from pasture."

This recovery time is much slower in meat products. Current research indicates that *the CLA in an animal that is ever fed grain never fully equals one fed no grain at all.*

In Utah, research animals that were fed small amounts of grain during their stocker stage and then finished on pasture alone had less than half the CLA of cattle completely grown from weaning to harvest on grass.

Dhiman said feeding an animal hay or wilted silage would decrease the CLA by one-third and green chop by 10%.

He said what created CLA were the highly volatile

---

**One glass of whole milk from a grazed cow plus one serving of grassfed beef or lamb per day would meet the CLA percentage required to provide effective cancer prevention.**
**Dr. Tilak Dhiman, Utah State University**

fatty acids in the grass that are quickly lost due to wilting.

Also fine chopping hay or silage decreases the CLA even further.

Feeding a ruminant animal fat also decreases CLA content.

He said the best way to increase the CLA content in meat and milk was to add a legume to the pasture.

"Clovers increase CLA content by 30% over straight grass pastures. Any forage that produces a higher average daily gain also produces a higher CLA percentage."

Breeds have an influence as well. He said the CLA is found within the animal's fat. Breeds that fatten easily on pasture are the breeds needed for high CLA products.

Supplements such as fish oil and plant oils can also increase the CLA in cows' milk but not as much in the beef.

"It is far easier to increase the CLA content in milk than in beef. We have found no CLA benefit in feeding oils to beef cattle."

To make this already excellent health food story even better, he said recent research has discovered a new fatty acid called Trans-vaccinic Acid or TVA. Human digestion can convert about 20% of this TVA to CLA.

He said this gives grassfed products an even further advantage because TVA has been found to increase linearly with the increase in pasture consumption just like CLA.

"The more we research grassfed products the bigger the (health) advantages to grass feeding gets," he said.

---

**Mice fed high CLA diets grew denser bones that were higher in calcium and phosphorous than the research controls. This suggests that a diet high in CLA foods — grassfed meat and dairy products — could help women with osteoporosis as well as breast cancer.**

**Dr. Tilak Dhiman, Utah State University**

And, here are even more health benefits.

■       Dhiman said Vitamin E is a potent anti-oxidant that lowers the risk of both heat disease and cancer. Anti-oxidants are also considered to be "anti-aging" agents that help your body resist the negative impacts of oxidation. He said Vitamin E is diet related and that grassfed beef has 300% more Vitamin E than grainfed beef.

■       Vitamin A is also a cancer fighter and is linked to both good vision and good sex. It is also necessary for bone development and the prevention of skin disorders. Grassfed beef has 400% more Vitamin A in the liver than grainfed beef.

■       Grassfed beef also has 75% more Omega-3 fatty acid than grainfed beef. Omega-3 is a potent anti-cancer agent and is essential for a sharp, well-functioning brain. It also reduces blood pressure and lowers heart attack risk. For good health a 2-to-1 Omega-6 to Omega-3 ratio is recommended. The diet of most Americans is currently estimated to be 20-to-1.

Jo Robinson, nutrition maven and co-author of *The Omega Diet*, said that most animal scientists don't seem to think there is a connection between what they feed animals and human health but there is.

"One, if it's in their feed, it's in our food. Two, if it's in our food, it's affecting our health. The new paradigm is we are what our animals eat," she said.

As previously pointed out, Americans currently have a gross imbalance of Omega-6 to Omega-3 fatty acids. Robinson said this is largely due to the feeding of wilted forages and grain to dairy cows and to the widespread practice of grain finishing beef.

She said research on people who had at least one heart attack and were put on a high Omega-3 diet showed a 70% decline in mortality. These same people also showed a 61% reduction in cancer deaths.

Omega-3 is an important brain fat. Children who were fed high Omega-3 diets had an IQ nine points higher than the average American child. High Omega-3 diets have also been found to be effective in treating depression.

She said grassfeeding not only increased the Omega-3 level in ruminants but also in the eggs of pastured poultry and the meat of pastured hogs as well.

***Currently the USA has the least amount of CLA in its diet of any country in the world.*** However, this is a relatively recent phenomenon.

Robinson said that in 1947 when she was born, 70% of all the beef eaten in the United States was from grassfed animals. "We don't need artificial designer food. We just need to get it back to where it was. We want to eat what is normal and natural."

She said Omega-3 fatty acid was extremely unstable and rapidly dissipated if exposed to air. This is why hay and wilted silages are low in Omega-3 fatty acids and cannot replicate the healthy meats and milk of direct-grazed pasture.

"You will never have a non-direct grazed beef product that is high in Omega-3 and CLA and Vitamin A and Beta Carotene."

She listed the following five points as the "whole story" of grassfed beef.

**1. The grassfed product is natural and wholesome with no artificial hormones, antibiotics or pesticides.**

**2. The grassfed product provides superior human nutrition. It is low in saturated bad fat, high in Omega-3, CLA and vitamins.**

**3. Grassfed animals are healthy and happy.**

**4. Grassfeeding not only is good for the environment but can actually improve the environment.**

**5. Grassfeeding is good for the survival of small farmers.**

Tilak Dhiman said Americans' reverence for science sometimes concerned him because it often blinded us to the inherent rightness of the natural world.

"Your best health will result from your eating as close to nature as possible. We are creatures of nature, not of lab science," he said.

One of the things working in the favor of all health-related businesses is demographics.

Business guru, Peter Drucker, said the surest predictor of the future is demographics. Every 50-year-old in the world already exists and can be counted. There will be no surprises as to their number. This creates a lot of predictability about the future.

The baby boom generation in the United States is the largest generation in history and currently makes up nearly one third of the entire population of the country. A similar generational bulge exists in Canada, Australia, the UK and New Zealand.

"The secret to wealth is to put your money where the baby boomers are putting theirs," said Robert Kiyosaki, author of the *Rich Dad/Poor Dad* book series.

This has been true since about 1950 and will be true for at least another 20 years.

I believe health-oriented food is going to be where baby boomers are going to put a lot of their money in the next few years.

Every eight and a half seconds an American boomer turns 50.

Currently the baby boom generation is nearing 50 as the mid-point in its aging progression. This means half are over 50 and half are under 50.

Those of you who have experienced a mid-life transition know that one thing that happens to you is that it is very easy to get overweight as your body metabolism slows down.

Have you noticed the sudden media interest in diet and other "weighty" issues? This is because the baby boomers are

---

**CLA is the only compound of animal origin that has shown in research trials all over the world to reduce cancer risk.**
**Dr. Tilak Dhiman, Utah State University**

interested in it. We are getting fat and we hate it.

Who do I sue for the way I look? McDonald's? Coca-Cola? Archers Daniel Midland?

Luckily for us meat producers, the late Doctor Adkins convinced a great many consumers and an increasing number of medical doctors that it is not the meat in the hamburger that makes you fat but what brackets it — the bread, the French fries and the soft drink.

A 20-year-old can eat junk food and get away with it. A 50-year-old can't.

While being overweight is certainly not fun, there are far worse things to come for aging boomers.

All of the medical breakthroughs of the 20th Century were in conquering the infectious diseases that killed the very young. Now the emphasis is shifting toward those diseases that manifest themselves later in life.

In one's late 50s, chronic illnesses such as arthritis, diabetes, cancer and heart disease start to surface. Currently, one out of two American males can expect to fight cancer and one out of every three females. Eighty percent of all cancers occur after age 55.

Coronary heart disease is the largest killer of both men and women. From age 50 to 60 is the first great winnowing of every generation due to heart disease.

Just wait until the leading age of the baby boom hits 60. The media will be screaming about the "sudden" health crisis in America.

The old 1960s joke "if I'd known I was going to live this long, I would have taken better care of myself when I was young" is not going to be funny for many boomers.

Health professionals are nearing agreement that any-

---

**You can defintely feel confident in promoting your grassfed products as a cancer preventer.**
**Dr. Tilak Dhiman, Utah State University**

where from 50 to 90% of these slow-to-develop chronic illnesses are diet and lifestyle related. The Center for Disease Control pegs it at 50%. Former Surgeon General, C. Everett Coop, pegs diet as the cause of 68% of chronic illness.

As the boomers age, the media fury will shift from today's overweight concerns to staying alive.

And by alive, I mean a quality life — not one wracked with pain and suffering. Unfortunately, we now have the technology to keep people alive but often in a shadowy painful hell somewhere between life and death.

Such a fate scares me to death. And, it scares most boomers.

What is hopeful (marketers pay attention!) is medical research showing that aging people can largely rejuvenate themselves through a combination of exercise and a nutritionally correct diet.

A common theme is that many of our health problems are being created by eating unnatural highly processed foods. White sugar, white flour, white rice and white fat on grain-finished animals have all been strongly implicated in obesity and increased heart disease.

The big story, of which grassfed products are a part, is that *eating as close to Nature's model as possible is the most healthy way to eat for rejuvenation.* Some people call this "beyond organic." I call it non-industrial food production.

---

**Green Beef Trend**

**Marsha McBride, executive chef and owner of Cafe Rouge in Berkeley, California, has received so many customer requests for grassfed beef that she gave in and bought some. With every article on healthier beef or the E. coli scare, McBride gets swamped with curious customers. Green beef, as it is called, is the latest trend in the food-obsessed Bay Area.**

**Contra Costa Times**

The mantra of the generation that grew up on "sex, drugs and rock and roll" is going to shift to one whose new orientation will be "live long, live healthy, die fast."

I predict this new mantra will hit its full flower when the peak of the baby boom hits 60 in a few years. That's when we really need our production prototype perfected and ready to roll out our story and product.

Among our biggest potential allies are the health-insurance companies. The one vendor you have in the health field that wants you to stay healthy is your health insurance company.

We are not far off from having health insurance rates

---

**Choice Grade Easy with Grassfed Steers**

Pueblo, Colorado, rancher, Russ Maytag, said he got his grassfed production ahead of his marketing in 2003 and still had a handful of two-year-old 1450-lb Red Angus steers on hand with winter coming on.

Through a friend with a small feedlot, he contacted National Beef about harvesting the steers. He said National Beef was extremely skeptical about the quality of a grassfed product and they would only do it on a grade and yield basis.

"The results were that every steer graded Choice and the fat color was also acceptable," Maytag said. "They said to send more just like them anytime.

"I was extremely skeptical when Allan Nation told me to take my heavy steers through another winter, but he was right. My cattle aren't fully grown until they weigh around 1450 pounds. And, cattle won't marble until they are fully grown."

Maytag said it takes him 26 to 27 months to grow the steers from birth to harvest. This is an average daily gain of 1.85 pounds for the entire period, which is competitive with Argentine graziers.

set by lifestyle and diet. Just as non-smokers get a discount today, people who exercise regularly and eat correctly will get one in the near future.

A major problem with today's health care system is that they profit when people are unhealthy. Consequently, there has been very little emphasis on preventative medicine.

We need to change the system so they profit primarily from creating health rather than treating disease. Already, in Seattle and South Florida a subscription-based form of health care where your doctor is paid a fixed annual stipend to keep you from getting sick is being tried.

Wouldn't it be great to have a doctor who had never had a sick patient? This needs to be the new medical paradigm.

The Internet is providing millions of aging boomers with alternative health information that is forcing mainstream health providers to run to keep up with the latest findings. There is a healthy skepticism about physicians today that is forcing them to run to keep up with their patients regarding the latest medical findings.

As one doctor said, "The pedestals are gone and, in terms of intellect and ability, aging boomers and their doctors stand on level ground."

Another thing we have going for us is the people we have attracted so far. The book *The Tipping Point* said that all national trends in America start in Marin County, California. Luckily that is exactly where the fire burns hottest for grass-finished beef today.

As far as media allies go, you can't beat *The New York Times* for making something new fashionable. So far, they have been a great friend to the whole non-industrial food movement.

You may not read *The New York Times* but I guarantee you that every magazine editor and television news producer does. What *The Times* decides is important soon becomes important because all the other news media pick it up and run with it as well.

We even have a smattering of support building on the

political front. A few politicians are realizing that a $100,000 income from a 100-acre farm could be just the ticket for economic revival in some of America's economically depressed rural areas.

As grass farmer, Joel Salatin said, "(Non-industrial food production) starts with the belief that how you produce your food really does matter."

There is another aspect that many pioneer grass finishers have commented to me about and that's the increased sense of spirituality that frequently accompanies a shift to producing and selling "real food."

Most of us become more spiritual as we age and come to grips with the idea of our own mortality. I know I certainly have.

I personally believe that all of us in this grass "movement," or whatever you want to call it, are doing something that is bigger and much more important than just building a business.

I believe a higher power programmed our food to naturally give us good health and we have lost sight of His handiwork.

Once you feel this way, I defy you not to feel at least a twinge of "awe" as you go about your work. Awe in this instance being defined as realizing you are in the presence of God.

# CHAPTER 7

# Three Proven Prototypes

Because of the climatic diversity of North America, no one prototype is going to fit everywhere. What I have found are three prototypes, which will pretty well cover most areas suited for grass finishing.

Of these three, Argentina's prototype probably has the greatest application for those wanting a year around program. New Zealand offers a good prototype for a seasonal, all-perennial system and Ireland for areas with exceptionally rainy winters that restrict grazing. Feel free to pick and choose elements from all three to fit your area and business plan.

## ARGENTINA

Argentina has an annual flow of grass-finished cattle from the pasture to the abattoir that is every bit as stable from one month to another as North America's feedlot system. It is located at the same latitudes as the United States and has the same "continental" climate of hot summers and cold winters.

However, *there is nothing "natural" about producing a year-round supply of grass-finished beef.* To do so requires a "whole" consisting of suitable soils, favorable climate, animal genetics and knowledge of how to plan and execute what the Argentines describe as "forage chains." These will be discussed in more detail in a subsequent chapter.

A year around supply of finished beef is not possible from

any one species of perennial pasture in continental climates like Argentina and the USA where annual temperatures can vary by over 100°F.

I have been to Argentina a half dozen times but I never fully understood it until I met Anibal Pordomingo.

Anibal comes from a ranching background, got his Ph.D. in the USA, worked at the Pampas Research Station and teaches at the University of La Pampas in Santa Rosa, Argentina. Santa Rosa, where he currently lives, is at the same latitude and rainfall as Tulsa, Oklahoma. The "finishing zone" that specializes in finishing beef cattle with grazed forages extends in USA terms from Dallas to Omaha, Nebraska, and in area is about equal to one half of Kansas and Oklahoma.

This small area not only finishes the majority of Argentina's 50 million head but also produces most of its export grains as well. Flat as a table-top, it is made up of wind-blown Loess soil and as such is highly mineralized.

Anibal said that *quality grass-finished beef starts with a fertile soil high in available calcium, sodium and phosphorous.* Such soils produce sweet meat and little bloat even on predominantly legume pastures.

Year-round finishing requires a high enough latitude to grow cool-season perennials and yet low enough for infrequent deep snow. Such a climate exists around the world one hundred miles North and South of the 38th degree of latitude in both the Northern and Southern Hemispheres.

This degree of latitude includes the Argentine Pampas, the Waikato of New Zealand, the Dandenong Finishing Zone of

---

**In Argentina, our grassfed beef supply has no more seasonality than is found with feedlot-finished beef in the United States. Argentine grassfed beef is known throughout the world for its excellent eating quality.**

**Dr. Anibal Pordomingo**

Australia, the Shenandoah Valley of Virginia, the Bluegrass region of Kentucky and the Flint Hills of Kansas. This climatic zone is generally referred to as the "Fescue Belt" in the eastern half of the United States.

On the West Coast, the 38th latitude runs just north of San Francisco. Anibal said the dry subtropics, such as in California lend themselves exceptionally well to an irrigated, alfalfa-based, beef-finishing system. He said that with irrigation, California's climate is far superior to Argentina's for grass finishing.

Non-irrigated finishing typically requires an annual rainfall of 35 inches and a sizable back-up of stored forages to overcome seasonal rainfall variations. The primary stored forage used in Argentina is alfalfa hay.

*Some Argentine ranchers have cow-calf ranches in the humid sub-tropic regions similar to Florida and finishing ranches in the cooler mid-latitudes.* Others have cowherds in the semi-arid foothills of the Andes.

The sub-tropic zones lend themselves well to late summer and fall calving and provide weaned calves to load up pastures at the end of winter in the temperate zone. Anibal said having access to both spring and fall calves allows for a better match of pasture and cattle than having only spring-born calves.

Most Argentine finishers prefer to produce their own calves rather than buy in outside calves for quality control. Anibal said it is important that the cattle be as genetically homogenous as possible for good eating quality. In a grass finishing

---

**On perennial forageses, spring finishing is usually the easiest and in many climates the only time grass finishing will be possible. However, if we will consider the use of winter and summer annuals, the possibilities for year-round production are greatly enhanced.**             **Dr. Anibal Pordomingo**

system utilizing both heifers and steers, the cattle naturally spread themselves out even with only one calving season.

There is typically no division between stocker graziers and grass finishers in Argentina. There are only cow-calf producers and stocker-finishers. As in the United States, the cow-calf producers who sell weaned calves tend to be part-time producers.

In 2003 Argentine ranchers were receiving only 30 cents a pound for their beef. Their agronomic costs of production were around 18 to 20 cents per pound of beef, which are similar to Oklahoma's.

While labor and land costs are slightly lower in Argentina, energy and fertilizer costs are double those of the USA. However, Anibal said *the big difference was that Argentinean ranchers use much less machinery* than the average American rancher even with their heavy use of annual pastures.

Most Pampas grass finishers produce 300 to 600 lbs of beef per acre with no nitrogen fertilizer, including those acres cut for hay or silage.

Argentine ranches are typically subdivided into large permanent paddocks with one wire electric fence. No gates are used as the fence is just raised with a special pole, called a vela, that has insulated handholds for the grazier but carries a current to keep the cattle from knocking it down.

Permanent paddocks are laid out so as to delineate common soil groups and similar topography. These paddocks are grouped into 10 paddock "modules."

---

**Winter annuals have the quality and the energy to sustain high weight gains in winter. Actually, the best weight gains available in grass-finished cattle have been accomplished in winter with winter annual grasses.**

**Dr. Anibal Pordomingo**

A typical ranch might have two to three 10-paddock modules. These modules are used to plan a shifting sequence of annual and perennial pastures called a forage chain.

These large permanent paddocks are further subdivided into small temporary paddocks with portable electric fences. Most stockwater is from windmills pumping to large above-ground reservoirs, which feed the water to smaller tanks.

Anibal said the overall stocking rate was the most critical factor on a finishing ranch. Finishing ranches typically have to have much lower stocking rates than those used only for stocker cattle and require a fall stocking rate half that of spring.

*He advocates a policy of stocking a ranch for the worst year rather than the average.*

"We always plan for the average and it never happens," he said.

"If you plan for the worst, the worst thing that can happen is you have too much grass. Not a bad problem."

Such a conservative stocking rate means no more than 400 lbs of animal per acre or one 1000 lb animal unit per hectare. A hectare is 2.5 acres.

Anibal said an 18-month-old, 1000 lb steer is the basic finished animal in Argentina. Heifers are harvested as yearlings at between 600 and 800 lbs. *It is this combination of harvest weights and sexes that helps give Argentina a year around harvest.*

The optimum time required to finish a six-month-old calf on pasture is 10 to 12 months.

---

**Properly managed, the winter annual will produce two to three times more forage than a cool-season perennial can in winter, and it can double the average daily gain available from the perennials.**

**Dr. Anibal Pordomingo**

Cows typically weigh between 850 and 950 lbs and are of Shorthorn, Angus (red and black) and Hereford breeding. Half of the nation's cows winter on corn stalks. The rest on frosted warm-season grasses and range. Feeding stored forages to grown beef cows is seldom if ever practiced.

Anibal said that pasture finishing was just as feed efficient as a feedlot if the correct genetics for pasture finishing are used. He described these *genetics as medium frame, early maturing, easy fattening cattle.* He said the Shorthorn/Angus and Hereford/Angus were the most common genetics used.

Many Argentine graziers have tried growthy North American cattle genetics with poor results. He said the most impossible animals to finish on grass are North American Holsteins.

Average daily gain is a product of forage type, quality and quantity. He said it was possible to have grazing animals gain 2.75 lbs a day every day of the year but that this was not the most economic utilization of the grass. There is no discernable difference in the quality of the meat between an average daily gain of two pounds a day and 2.75 lbs.

He said *it is always more efficient to concentrate on raising one's worst gains of the year than in trying to better the best.* Frequent weighing of the cattle is necessary to know what is going on, where and when, in your operation. Cattle should always be harvested when they are gaining well for the tenderest meat.

Interestingly, Argentine beef is aged for four to seven days. He said there were no consumer concerns about beef tenderness in Argentina but frequent complaints about the meat being too fat. He said marbling was not a major concern because Argentineans cook lean meats slow.

### Leader-follower Grazing

Anibal said Argentine herds are divided into leader-follower groups.

The leader group are the animals the closest to being

finished. The follower group are what are known as stocker cattle in North America and are growing frame rather than fattening.

The heavier the steer becomes the higher the quality of pasture he requires. However, the heavy steer will also show high gain response to an increase in pasture quality. Lighter cattle can do better on lesser quality.

A major problem with novice graziers is that the follower stocker cattle's gains are often allowed to drop too low due to poor pasture allocation. As a result, *Anibal recommends that dairy quality free choice alfalfa hay be kept available to these follower cattle at all times* or to use cows and calves as the follower herd.

"Often times if you don't have the management skills leader-follower requires it is better to just mow it down," he said.

The secret to high average daily gains is found in the "easiness" of the grazing. The leader herd should be able to eat up to three percent of its bodyweight in dry matter with no stress or effort. In other words, finishing cattle should only top graze the very best of the pasture.

The finishing cattle should be shifted at least once a day but twice a day is better. The stocker cattle can be shifted every three days.

While theoretically the grazier should leave 35% of the pasture on offer, Anibal said it was better to take only half to give yourself some "wiggle room" to make mistakes.

While most graziers pay the most attention to the finishing group, the best increase in overall efficiency is to increase the gains of the stocker herd rather than trying to produce spectacular gains in the leader group.

*Herds should be limited to 200 to 250 head for maximum gains and low stress*. Plentiful stockwater is necessary for maximum gains. He said limiting stockwater to cattle on lush early spring or fall pastures in order to try to increase gains would actually worsen them.

"It takes lots of water to maximize gains."

If grass becomes short, he said to graze in the afternoon when carbohydrates are highest and feed stored forages off the pasture the rest of the day. Make sure stored forages are always available when grass is less than plentiful. Never, ever, let your stocker or finishing cattle get hungry, he said.

*Cattle should have access to pasture or stored forages 24 hours a day to avoid hunger and to balance their dry matter needs.*

### Summer Is the Most Difficult Season

In the mid-latitude region, Anibal said summer was the most difficult time to finish cattle. In hot summer areas, finishing cattle required the planting of highly digestible annuals such as corn and soybeans.

These annuals are necessary because cool-season

---

**Leader Follower Grazing Recommended**

**Grazing management is important in producing high rates of gain. For example, leaving a one inch higher residual in the paddock would increase average daily gain by 37%.**

However, leaving such a high residual produces a stemmy, unpalatable sward for the next round as there will be sharp stems poking the animals' noses. Consumption drops on subsequent shifts. To prevent this problem, grass-finishing systems need to be leader-follower grazing systems — animals nearing finish are allowed to "cream" the best of the forage leaving a high residual. Then second-grazers come in to take the forage to three inches.

Noble Foundation research found that leader-follower systems could produce 40% more gain per acre than single grazers.  ADGs of as much as four

*continued next page....*

perennial grasses and alfalfa typically produced only a half pound a day of gain in very hot weather (over 86° F).

*The benchmark for a "finishing" forage is the ability to produce an average daily gain of at least 1.7 lbs per day.* At gains of less than this the animals grow frame rather than fatten. For the best eating quality meat, calves should never be allowed to gain less than a half pound per day. The higher the overall average daily gain is, the more tender the meat will be.

High average daily gains are produced by forages with a digestibility of 65% or greater; a daily dry matter forage allowance of at least four percent of the animal's bodyweight; a forage dry matter of at least 20%; and a forage supply balanced in protein and energy.

Argentina finishers average 1.32 lbs for the year. *If cattle ever lose weight during their stocker-finishing period they will eat tough.* This is why high quality stored

---

*...continued*

**pounds a day had been produced by allowing first-grazer animals maximum grass selectivity.**

**A 3-grazer system could be a small group of finishing steers, followed by a larger group of growing steers, followed by a small group of dry cows or mature bulls. The third grazer will have slim pickings and needs to be a non-production class of animal.**

**In some wet, temperate climates such as Ireland and New Zealand only first grazers are used and the grass is kept in a relatively narrow three- to six-inch-height range. Such a system is unsuitable for the USA due to its highly unreliable rainfall.**

**In central Oklahoma using a leader-follower system, an initial grass height of around 10 inches produced both the best gains and the most beef gain per acre.**

**R.L. Dalrymple, (retired), Noble Foundation**

forages such as alfalfa hay must be kept available.

Anibal said care should be taken with vaccinations and wormers as these were stressful to the cattle and could affect meat tenderness.

Interestingly, he said *the most cost-effective time to supplement pasture-finished cattle with stored forages in hot climates is during the summer rather than the winter.* More on this later.

Until recently, the Argentine recommendation was to supplement low energy early fall pastures with grain to try to increase average daily gains. While carbohydrate supplementation looks good on paper and in research trials with precise hand feeding, Anibal said in actual pasture trials, grain, molasses and other carbohydrate supplements produced lower gains than un-supplemented pasture at least half the time.

"The problem with any such carbohydrate supplementation is that it must be fed at a very low amount to keep the rumen neutral. The precision required is just not practical for a pasture situation," he said. "I now say that if you can't do it with a forage, don't do it."

### Stored Forages must Be of the Highest Quality

Stored forages are used with finishing cattle to increase the forage supply, increase dry matter content of early spring and fall pastures, and to help balance the too high protein found in fall pastures by diluting crude protein in immature pastures.

> **Our research in central Argentina has shown that the use of hays (alfalfa, mixed pasture or oats hay) with at least 60% digestibility, or silages (alfalfa-based silage, mixed pasture silage, oats silage or soybean silage) with at least 65% digestibility, will not decrease the finishing gain on winter annuals even when it makes up 50% of the animal's daily intake.**
>
> **Dr. Anibal Pordomingo**

*It is critical that stored forages fed on pasture be of higher quality than the pasture otherwise they will lower the average daily gain.* Finishing animals are allowed access to free-choice alfalfa hay for up to 10 months of the year. The two months skipped are the late spring and early summer period.

Since no stored forage is of higher quality than pasture at this time of the year no stored forages should be fed. All stored forages should be analyzed for quality before being fed to finishing cattle and should be of dairy quality.

On lush, watery, fall and early spring pastures, allowing cattle access to spring-cut alfalfa hay can double average daily gains. However, for this to happen the cattle have to eat one percent of their bodyweight in hay. They will only do this if the hay is of the highest quality.

Spring-cut, annual ryegrass silage is best used in the summer to increase the average daily gain of the growing stocker cattle (who are not on corn or soybeans) rather than as a winter supplement. Direct-cut, vacuum silage is the best because it loses less of its carbohydrate portion because wilting is not necessary with this method. Annual ryegrass makes the best silage because it is higher in carbohydrates than any other small grain or cool-season grass.

Anibal said it was always better to buy-in forage tested alfalfa hay because haying rapidly thinned the stands and opened them up to weed encroachment.

---

**It has been suggested that on lush forages with a dry matter of less than 18%, the addition of fibrous materials to the diet would help to reduce watery feces and improve gains.**

**Although this will help reduce scouring, there is no evidence of true gain improvement from feeding low-quality hays. On the other hand, feeding high quality hays will increase gains.**

**Dr. Anibal Pordomingo**

## Plant Grass and Legumes Separately

Anibal said the Argentine "forage chain" starts with a base pasture of alfalfa and a companion cool-season grass. White clover could be substituted for alfalfa in more humid regions (40 inches and higher) and is actually a better forage than alfalfa because it does not lignify (become less digestible) in hot weather as alfalfa does. ***With all legumes, keeping soil sodium levels adequate is important in preventing bloat.***

He said the absolutely best perennial finishing pasture would be one of perennial ryegrass and white clover. However, such a pasture is hard to grow in continental climates like Argentina and the USA with their very hot and dry summers. Alfalfa should always be used in areas with less than 40 inches of rainfall as it is the most drought-tolerant legume.

A typical Argentine finishing pasture would be drilled to two rows of alfalfa and one row of grass or one to one. The grass and legumes are drilled separately so that the cattle can easily choose between the grass and the legume when grazing. Typically they will eat legumes in the morning and grass in the afternoon as this is when the grass is highest in carbohydrate, which the animal will be craving after a morning of high-protein grazing. For this reason, ***shifting cattle in the late afternoon to a new paddock is a good way to minimize legume bloat.*** I'll have more on this in a minute.

---

**In Argentina, the greatest individual gains on grass finishing programs are reached in mid to late winter on winter annuals such as cereal rye, oats, triticale or wheat.**

**Under restricted availability (>5% of bodyweight of available forage on dry matter basis/head/day) weight gains are usually from 2.0 to 2.7 lbs/day and feed efficiencies range from 9 to 14.1 (pound of grass dry matter per pound of weight gain).**

**Dr. Anibal Pordomingo**

---

The companion grass is included to lengthen the green season, utilize the excess nitrogen of the legume, to increase soil structure and lower legume bloat. Other than these attributes, the companion grass is not that important in producing gain. The majority of the gain comes from the legume. Consequently, Anibal said the pasture should be managed just like a pure stand of alfalfa. In other words, *it is the legume that is the key indicator plant to watch and manage for.*

Ideally the animals will get no more than 20 to 30% of their total dry matter from the perennial grasses. With alfalfa, the stand will thin with time. Once the alfalfa percentage falls below 50%, the pasture should be plowed down for a rotation through winter and summer annuals. This plow down period is an excellent time to deeply incorporate lime in acid soils.

Permanent pastures should never be planted back-to-back but should always go through a period of annuals first. Old permanent pasture grass is very resilient and will dominate a new alfalfa planting.

He said *giving animals as much choice as possible in what they grazed was a key element in producing a high average daily gain.* Consequently, overseeding of legumes into grass is never used in Argentine. All forages are drilled into separate rows that facilitage animal choice.

## Graze Legumes in the Morning, Grasses in the Afternoon

Animals typically eat legumes in the morning and shift to grass in the afternoon when the carbohydrate levels are at their peak. The carbohydrate in the grass will help the animals digest the high protein legume. *This daily shift in preference is a key in preventing bloat on high legume pastures.*

Animals on legume-dominant pastures should always be shifted to a new paddock in the afternoon when they are primarily eating grass. Animals on alfalfa should not be shifted in the morning particularly if they are hungry.

Also, once animals are on alfalfa-based pastures they should stay on them rather than being rotated to other pastures and back as it would require a rumen transition to digest the alfalfa. Animals' rumens can be prepared for going to an alfalfa pasture by feeding alfalfa hay free choice.

Soils that are high in phosphorous and sodium produce less legume bloat. *The best way to produce soils high in phosphorous is to lime the pasture* as without calcium, phosphorous is largely unavailable to the plant.

The higher the forage quality is of the companion grass the less problem legume bloat will be. The best companion grass is perennial ryegrass, but this species requires a very fertile, high-organic-matter soil to persist. His second choice would be orchardgrass, but it also is hard to keep in hot, semi-dry environments, whereas brome isn't. His last choice would be fungus-free fescue.

Only use fescue if absolutely nothing else will work in your area, and never use endophyte-infected fescue as it produces a highly objectionable off flavor in the meat.

Cool-season perennials mix well with alfalfa because they have a similar maturity curve with both responding well to the 30- to 35-day rest period alfalfa requires.

Alfalfa varieties bred for grazing are the best to use and non-dormant varieties are preferred in the warmer winter regions. Regrowth can be stimulated by grazing as little as one-third of the alfalfa plant. Trying to replicate the look of a hay

---

**For most improved pastures of central Argentina (temperate subhumid region similar to Central Oklahoma), paddock size and stocking rate are set on the basis of a daily dry matter intake of 3% of the animals' bodyweight (for growing-finishing beef cattle), so that the resulting grazing pressure will remove the 50 to 60% of available forage in 1 to 7 days and pastures will rest for 30 to 45 days (period that depends on re-growth rate).**

**Dr. Anibal Pordomingo**

cutting is not the way to produce high animal performance or long-lived stands. This is particularly true in sunny hot climates where the plant crown can be damaged by heat.

The alfalfa stand typically starts to dramatically decline after four years of grazing and the permanent pastures should be plowed down and planted to a planned sequence of grazable annual forages.

The plowed down permanent pasture typically produces enough soil nitrogen to grow three years of grazed annual crops. Because in a grazing situation most of the nitrogen is recycled in the manure and urine, very little (7 to 8%) of it escapes the system in the product.

Anibal said it is the lack of need for outside fertility that makes grazing financially superior to cropping.

*"With grazing the soil gets better and better. With cropping, it gets worse and worse,"* he said.

### Artificial Nitrogen Lowers Daily Gains

He said artificial nitrogen should be used sparingly if at all as it can actually lower average daily gains of fattening cattle by making the forages too high in protein.

He said to keep in mind that *it is carbohydrate that produces fat and not protein.*

Forages in excess of 15% protein require energy to convert the excess nitrogen to urea and excrete it. This energy comes from the animal's fat reserves and works against producing a well-finished animal.

Given the opportunity to choose between grasses, legumes and a high-quality stored forage, the animal can pretty well balance his protein, energy and dry matter and produce maximum weight gains.

Keep in mind that *it takes time to build the rumen bugs necessary to digest different forages. Feeding hay or silage of the next pasture forage prior to going to that forage can help develop these new bugs and prevent the low gains common to a forage switch.*

### Winter Annuals Should Always Be Planted Separately

For winter grazing, cereal rye, triticale, wheat and forage oats are used. Argentina's forage oats are bred to be very leafy and cold-tolerant and can be used similar to winter wheat. Annual ryegrass with its slow maturity and high carbohydrate content is particularly prized.

After plowing down the alfalfa/grass permanent pasture, Anibal said to *allow for a digestion period for the soil microbes to break down and release the pasture's plant nitrogen* into the soil for subsequent use. This digestion period for plowed down permanent pasture is 45 days and winter annuals require 30 days.

He said that no-till could be used for all subsequent forage plantings but that it should not be used to remove the permanent pasture due to accumulated soil compaction. A deep ripping plow is best to break compaction and to incorporate lime deep into the soil to alleviate sub-soil acidity, which will prevent deep-rooting by the alfalfa.

*A key element in a forage chain is to plant these small grains in separate paddocks rather than mix them together. This is because they have different maturity periods.*

For example, cereal rye grows at the coldest temperatures but matures very early in the spring. Triticale matures slightly later and wheat and oats even later. The cattle being finished are shifted to the forage most in its prime and as the forages mature they are plowed down and planted to a summer annual after a 30-day "digestion" period.

This staggered plowing and planting prevents the forages

---

**If soil allows, the need for establishing new pastures on a 5-year frequency may not be a disadvantage. It gives the opportunity of incorporating genetic improvements and the possibility of land rotation with annuals (winter annual pastures or annual crops).**

**Dr. Anibal Pordomingo**

from maturing all at the same time and is a key element in the "chain." It also allows a very large finishing ranch to cover a lot of ground with a minimal amount of machinery

Winter annuals are always planted with companion legumes. The two most common are red clover and yellow trefoil. These annual legumes are included more for increased forage quality and nutrition balance than for their soil-nitrogen-fixing ability.

The cattle should be shifted to permanent pasture and stored forages to prevent soil pugging on the winter annual pastures during high rainfall events. However, the cattle should be allowed to graze a few hours each day on the annuals to keep their rumen bugs functioning.

*The biggest problem Anibal has seen in America is that many graziers turn their cattle onto winter annuals when it is too immature.* Such young forage is very high in water content and nutritionally imbalanced and will produce gains of less than a half pound a day.

Small grains should not be grazed until they have reached six to eight inches in height and have a dry matter content of 18 to 20%. If the dry matter is lower than 18%, feed high quality alfalfa hay free choice.

On the first pass, Anibal said to only top graze the annuals. This will allow for faster regrowth and more grazings.

Once the weather becomes cold, winter annuals' quality improves as nutrients become balanced and dry matter increases. He said you will actually be able to produce your highest gains in the winter. *Winter finishing of cattle is only a matter of early planting and subsequent allocating of forage.*

If quantity becomes limiting, graze for five or six hours in the afternoon and feed high quality alfalfa hay or spring-cut pasture silage off the pasture the rest of the day.

Anibal said that wheat, rye and triticale all flavored the meat but that Argentineans didn't find this objectionable because they were used to it. This off flavoring can be reduced with the addition of legumes such as vetch, which will also extend the period of high average daily gains.

*Annual ryegrass produces the best flavored meat of all forages because it offers the best balance of protein and energy.*

Winter annuals are sometimes drilled in a cross-hatch manner into fall planted stands of alfalfa to provide winter grazing.

### Soybeans Good Mid-summer Finishing Forage

Soybeans are very drought tolerant and provide 45 to 60 days of grazing in the mid- to late summer.

Graziers should choose the longest maturity variety of soybean they can grow in their area for grazing. This will allow two to four grazings 25 to 30 days apart before seed pods appear. *Grazing must be discontinued when the beans appear as they can cause severe health problems for the cattle.*

The cattle should only graze the green leaves and not the stems for high average daily gains (two pounds). There are no bloat problems with soybeans. *The cattle should be shifted when 50 to 60% of the leaves are removed.*

Plowing down green soybeans at the onset of seedset is an excellent way to improve soil fertility. This is the equivalent of adding 100 lbs of urea per acre.

The best system for maximum gains would be to allow the cattle to graze soybeans in the morning and green corn in the afternoon. In this way the high carbohydrate corn will help the animal digest the high protein soybeans.

---

**Forage soybeans combine well with corn in hourly strip grazings. Corn provides more sugars and dry matter, and soybeans complement with protein and digestible fiber.**

**Forage soybeans combine well with energy supplements such as molasses.**

**Dr. Anibal Pordomingo**

---

## Green Leaf Corn for Grazing

Corn for grazing should be planted in three plantings 15 days apart. Twice as much seed should be planted in the row but the rows should be the same width as for corn for harvest.

Open pollinated corn is much more drought tolerant than the hybrids and is more digestible. Silage corn varieties are better than grain varieties for grazing.

There is no negative rumen adjustment from alfalfa pasture to green corn.

***The cattle should enter the corn paddock when it is at your shoulder height and only eat the leaves of the plant.*** Forcing the animals to eat the lower half of the stalk will tremendously lower average daily gains. This remnant can be grazed with cows.

Cattle that have grazed corn leaves should be harvested and not returned to permanent pasture as they will lose weight and this will toughen the meat.

## Grass and Grain Production Together

Many Argentine pampas graziers also grow grain crops. Integrating grain production into the forage chain greatly complicates things but can add marketing options.

Rotating land through pasture can greatly increase subsequent grain yields by providing soil organic matter and thereby would allow high-value organic grain production.

Anibal said that grazing legumes was virtually the only way to bring a cropped-out farm back to life. However, this should not be attempted by the impatient nor the poor as it required at least eight to ten years to rejuvenate the soils of a cropped-out farm.

## A Working Example

To give you some idea of how "scalable" grassfed beef production can be, you should visit Eduardo Pereda's Nueva Castilla Ranch in La Pampa Province. I have been to his ranch three times and he has always welcomed me and has actually opened his books to show me that even at Argentina's low beef price he was earning a 17 to 18% return per annum.

On my last visit Pereda had some 23,173 head of finishing cattle on his ranch's 36,300 acres. Half of these cattle were from his cow-calf ranch in the brushy Andean foothills to the west and half were purchased from the sub-tropic zone near Paraguay.

Pereda has 10,000 cows on the cow-calf ranch and maintains another 3000 mother cows on the finishing ranch as "grass scavengers." These cows are used to recycle crop residues and to clean up roadsides, swampy areas and conservation plantings.

Due to the arid nature of his cow-calf ranch his breed choice is Hereford. He said he preferred small cows because of their greater efficiency and faster finishing. He said his average steer harvest weight was 900 lbs on his home-raised cattle and 968 lbs on the purchased cattle.

He said to harvest the animals at such a light weight required that they be overwintered at a high rate of gain. He said steers that are allowed to grow slowly over winter grew too much frame to finish in the spring. To keep winter gains high he uses small grains with the finishing animals and feeds silage under the wire for the stocker class of animals.

He said his primary management goal was to fully utilize his high quality late spring and early summer forages. He said his ranch grew 40% of its total dry matter in just 90 days despite an extensive "forage chain" of annuals. It was for this reason, and this reason only, that he purchased outside cattle.

---

**Good hay at all ltimes on the field is a safety factor to help counteract an allocation error. Never let the cattle go hungry. The last thing we want is the pasture destroyed by overgrazing.**

**Dr. Anibal Pordomingo**

***His goal is to produce 500 lbs of gain per acre on Nueva Castilla.*** His direct cost of gain (before fixed overheads) is 18 cents per pound.

The ranch is divided into 110 large permanent paddocks with an average size of 330 acres. These large paddocks facilitate occasional cropping and are further subdivided by one-wire electric interior fences.

There are no gates on the ranch. The wire is simply lifted with a ten foot "hot" pole called a vela (candle) wherever a gate is required. All ranch vehicles are fitted with permanent bottom rails and a removable V nose attachment (Think of an upside down locomotive cow-catcher) that allows driving over all of the ranch's fences at 30 mph. He said such drive-over fencing greatly improves management productivity on large properties.

An example of this emphasis on labor productivity is that the huge ranch only requires 17 cowboys. Pereda said their primary job is to shift the cattle to fresh pasture every three to five days.

The ranch also grows corn, wheat, rye and soybeans. All of these crops can be grazed for beef or harvested for grain depending upon the price. He said it took a very high grain price to bid his corn away from direct grazing due to the much lower harvest cost.

To give you some idea of the capital lowering consequences of staggered planting and plant species diversity, *the 30,000-acre Nueva Castilla Ranch near Santa Rosa has only six tractors, three planters and three grain drills.*

The ranch plants all of their crops using their own machinery but any grain harvesting is done exclusively by contractors. This lack of investment in harvesting machinery not only keeps captial costs low but makes the direct-graze option more attractive.

Pereda said the diversified crop and cattle mix provided excellent year-around cash flow, steady work for his labor force and better equipment utilization.

All planting is done no-till as his ranch is very susceptible to soil wind erosion. He said they planted 21,600 acres of grain and pasture crops in the spring and 10,800 in the fall.

He said the decision on whether to machine harvest or graze out a crop was made depending upon the price of grain. However, he far preferred to direct-graze his corn and small grain crops.

Anibal said without the direct-graze option the market risk of grain farming would be too high to consider.

### Argentine Finishing Tips

■        Soils high in calcium and phosphorous produce sweet meat and little bloat.

■        Use cattle as genetically homogenous as possible.

■        Stock your ranch for the worst year, not the average.

■        Concentrate on raising your worst gains of the year rather than your best.

■        For tender meat always harvest your cattle when they are gaining well.

■        Never, ever, let your cattle become hungry.

■        Only use forages capable of producing an ADG of 1.8 lbs per day.

■        Stored forages must be higher in quality than the pasture they are fed on.

■        Drill grass and legumes in separate rows to give cattle choice.

■        Only shift cattle on legume pastures in the afternoon.

■        Nitrogen fertilizer can lower your cattle's gains.

■        Always plant different forage species in separate paddocks rather than mixed together.

■        Cattle that have grazed corn should be harvested and not returned to pasture.

■        Soybeans should be grazed in the morning and corn in the afternoon.

# NEW ZEALAND

The majority of New Zealand's domestic table beef production is from cull dairy cows (tenderloin) and two-year-old dairy bulls (tenderloin and ribeye) that are primarily harvested for export hamburger. Since export bulls typically bring the same price per pound as domestically consumed table grade steers and require no fat finish, they are the class of choice for most graziers. Unlike Argentina, which sells almost all of its beef domestically there is no cultural emphasis on beef quality in New Zealand; and lamb is the traditional meat.

As an export driven economy, there is a great emphasis on least-cost production methods, which is accomplished through seasonal production in tune with the annual forage cycle. As a result, New Zealand beef production is much more seasonal than Argentina's. They are also masters at stockpiling and rationing out fall grown pasture over the winter with strip grazing.

The Kiwi's preferred breed choice for table grade beef is an Angus/Hereford cross with a mature slaughter weight of around 1100 lbs. On average, this type of animal gains around 453 lbs in a 245 day green season graze or 1.8 lbs per day. Winter gains on stockpiled cool-season pasture with no supplementation average 158 lbs in 120 days for an average daily gain of 1.3 lbs.

A higher winter average gain can be achieved by supplementing with winter annuals, brassicas, or pasture silage.

Increased gains for all of these inputs are virtually identical with a 10-to-1 ratio of input to output. In other words, two pounds of supplementary brassica dry matter increased average daily gain by two tenths of a pound of gain.

However, at New Zealand's low beef prices, artificially boosting winter gains was said to only be cost-effective when used to match cattle growth to a targeted (contracted) premium-priced, out-of-season slaughter date. In New Zealand and in all grassfed countries I have visited, late winter and early spring animals are the highest priced of the year.

Pasture finishing tends to be concentrated on fertile soils as these produce a more even year-round forage growth curve thanks to better summer moisture retention and a much better legume component. White clover is considered essential to both smooth the summer drop in cool-season grass production and to keep summer gains reasonably high.

*Soils low in fertility have a much more pronounced annual forage curve* and are best used with seasonal ewe-lamb and cow-calf programs that time their calving and lambing to the surplus period, and sell off the progeny in late summer and autumn.

Brassicas (turnips, rape, kale), which are preferred New Zealand supplemental forages with finishing cattle, need soils that are high in phosphate for economic production. Brassicas are 80% digestible and are capable of producing gains of 1.8 lbs per day in either winter or summer in two-year-old beeves.

Flatter lands also lend themselves better to machine harvest of excess spring forages. Hay is not considered a "finishing feed." Pasture silage is the preferred stored supplement due to its higher quality and lower machinery cost.

Heavy two-year-old cattle have been found to severely pug upland pastures so the lighter yearlings tend to be grazed in the uplands and the heavy cattle kept on the flatter lands.

On upland pastures where machine harvest is not possible, *a combination of sheep and cattle is considered absolutely essential to maintain the pasture quality high enough for beef finishing.*

Beef cows are also useful to keep the spring growth

> **New Zealand's beef production is very seasonal due to its emphasis on least-cost production. The harvest infrastructure survives by harvesting seasonal waves of veal calves, culled cows, export bulls and domestic table grade beeves. Lambs are harvested in the summer and early fall in the same abattoirs.**

under control as they can increase their grass consumption at calving to a much greater degree than can non-lactating cattle. New Zealanders said the maximum intake of dry stock is only 70% of lactating animals of a similar weight.

Perennial ryegrass is the forage of choice on high-organic-matter soils, and fescue and orchard on the drier na-tured, low-organic-matter soils. The New Zealanders said that perennial ryegrass needs a soil organic matter of nearly five percent to persist in their hot (80°F) dry summers.

The New Zealanders consider their unfair advantage to be their skill in stockpiling autumn pasture and tightly rationing it out so as to minimize stored forage supplementation.

With a national obsession with low cost production, *the New Zealand beef harvest is highly seasonal with much of it occurring in the summer period.* It can do this because New Zealand's beef production is largely export oriented.

At the heart of the Kiwi system of grassfed beef produc-tion is a seasonal variation in stocking rate that maximizes the use of spring pasture and then decreases feed demand through-out the summer in concert with the slump in summer pasture production.

Fall stocking rates are kept low to build stockpiled pasture for winter use.

A typical seasonal plan would be to sort off and harvest one-third of the animals in (the North American equivalent of) June, one-third in July and one-third in August. Any stragglers would be finished on brassica-supplemented, autumn pasture before winter. The age of the steers in this system when har-vested varies from 24 to 30 months.

This gradual summer destocking allows forage demand to follow the natural cool-season grass growth curve of peaking in the spring and bottoming out in late summer before a fall recovery.

The Kiwis repeatedly emphasize that keeping the stock-ing rate light is very important with heavy, finish-weight cattle. Here's why:

The slower a heavy animal is gaining the more of its feed is being used to keep it breathing. This is called body maintenance.

Greater feed efficiency occurs at higher feed intakes because maintenance requirements are satisfied before growth. The greater the feed intake the higher the overall efficiency of gain. Therefore, *it is more grass efficient with heavy cattle in the long term to use a lower stocking rate which gives you a higher average daily gain than a higher stocking rate that gives you a lower one.*

For example, a 650-lb animal gaining 3.3 lbs per day will have a feed conversion efficiency twice that of a similar weight animal gaining only a half pound a day.

Feed intake is directly related to digestibility. Perennial grasses are not digestible enough above 86°F to produce high rates of gain. This is offset in New Zealand by good summer growth of white clover. White clover is unique in that it is the only perennial forage plant that does not lose digestibility in hot weather.

Lighter, growing, stocker animals with lower maintenance requirements are much more grass efficient than heavier finishing animals because more of their feed is channeled to growth. Lighter cattle also have a much greater propensity for compensatory gain than heavier cattle because most of their gain is, in effect, water; whereas, with heavy cattle it is hard to produce fat.

This is why grassfed countries typically have their slaughter animals as the highest priced weight category. Heavy cattle

---

**We tried a three-herd, leader-follower system and found it was just too hard on the pastures. Now, we just use the cows to clean up after the finishing animals as needed. The cows calve in mid-March during the annual range's green season.**

**Mac & Kate Magruder, Potter Valley, California**

produce less beef per acre than yearling cattle and so require a price premium per pound to make them competitive in profitability with the alternative of grazing stocker cattle.

After having just told you that unrestricted intake is the best policy for finishing weight animals, I must quickly add that there is an exception.

Annual conversion efficiency is not influenced by winter feeding levels. This is because animals that gain less weight during winter have lower maintenance requirements in spring than heavier animals because they are smaller. This allows for a more rapid spring growth rate.

Given similar intakes in spring, lighter animals funnel more of it to growth. The result of this compensation is that the annual feed conversion is about the same irrespective of animal growth rate during winter.

Compensatory gain makes supplementation in winter less cost effective than it appears because one-third of the extra weight gain will be lost by the animal not gaining as well as it otherwise would have gained. However, animals must be at a weight at the start of spring where they will hit their "finish" weight before the worst heat of summer. The New Zealanders accomplish this balancing act by the frequent weighing of cattle.

Now, keep in mind that *the goal is to still have the animals gain during the winter. Actual weight loss produces tough beef.* Their point is that it may not pay to have them gallop through the winter.

This management of a growing set of yearlings and a finishing set of two-year-olds really separates the men from the boys.

While a leader-follower system sounds relatively easy, I should warn you that very few graziers are able to successfully pull it off. Most New Zealanders wind up running the two different classes as separate operations with their own separate sets of paddocks.

Since the yearlings are not as grass-management sensitive as the finishing cattle, many grass finishers prefer to put them out with custom graziers and concentrate their management efforts on finishing cattle.

## Out-of-Season Production

For out-of-season production, the New Zealanders devised a program whereby the heaviest 25% of the two-year-olds would be sold in (a Northern Hemisphere equivalent of) March and the remaining 75% sold in December, January and February. Note that the Kiwis skip the very difficult months of October and November and April when dry matter and plant energy are low.

The problem with out-of-season production is that, as you can see in the Out-of-Season graph from the New Zealand Beef Council, it requires a much flatter feed demand curve. The only way to achieve this is to run a much lower stocking rate than with a seasonal program. New Zealand research found that the overall stocking rate had to be reduced to 77% of the

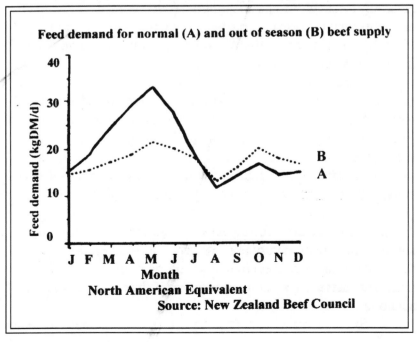

Feed demand for normal (A) and out of season (B) beef supply

Source: New Zealand Beef Council

normal stocking rate to stockpile enough autumn grass for winter finishing.

The general rule of thumb the Kiwis use is that *an animal in its final finishing stage should have no restrictions on its input.* Such a non-restrictive policy requires a much higher amount of stockpiled pasture than a carefully rationed winter program with summer-finished beeves.

This stocking rate is then too low to utilize the spring pasture lush and 30% of the farm has to be cut for silage and ten percent has to be shredded with a rotary mower at the end of May just to maintain quality. (Remember, all months are presented in their Northern Hemisphere equivalent.)

The net result of this lower stocking rate is a decrease in profitability of 20% over the seasonal program. In other words, the price schedule for winter and early spring production has to be at least 20% higher to break even with the normal seasonal system.

Dr. Gavin Sheath of Whatawhata said that systems such as this out-of-season finishing system that produce large spring-season grass surpluses lend themselves particularly well to the addition of a late spring calving beef cow herd.

"Breeding cow herds are very complementary to pasture finishing," he said.

*"The trick is to time calving to the oversupply of grass. In this way the cow complements rather than competes."*

As previously pointed out, the New Zealand dairy herd also supplies most of the country's beef production — 58% of which is exported to the USA as ground beef. Still there are estimated to be around four million beef animals in the country.

Again, unlike Argentina, New Zealand's beef production is very seasonal due to its emphasis on least-cost production. The harvest infrastructure survives by harvesting seasonal waves of veal calves, culled cows, export bulls and domestic table grade beeves. Lambs are harvested in the summer and early fall in the same abattoirs.

Export beef is primarily from grassfed Friesian bulls. The

extremely lean meat from these animals goes to American fast food restaurant chains who blend it with fat trimmings from grainfed beeves to make a fat and flavor-consistent hamburger.

The tenderloins and loins are pulled from these bulls for both domestic and export table grade beef sales. Like most of New Zealand's pasture-based production, this export bull meat is highly seasonal and is largely harvested in the Southern Hemisphere's late spring and early summer. This meat arrives in the USA in the Northern Hemisphere's winter when the domestic grassfed supply is at its yearly low.

While bull beef is definitely not gourmet, I thought you might enjoy a peak at its production nuances.

### A Working Example

One bull beef grazier I have kept up with over my many visits to New Zealand in the last 25 years is Eric Anderson who ranches near Huntley on the North Island.

Anderson leases 3000 acres, which he has divided into 320 permanent paddocks. These are further subdivided into 500 paddocks in winter. *These high paddock numbers are a necessity when grazing intact bulls as they must be kept in herds of no more than 20 to prevent fighting.* Therefore, each group of 20 bulls must have its own "cell" of paddocks. Hence, the high paddock count.

When I last visited, Anderson had 903 two-year-old bulls on the property. He grazes bulls for only eight months of the year to allow for fall stockpiling of his pasture for wintergrazing. He said winter grass was worth three times as much as late summer and fall grass in New Zealand.

---

**Seasonal variation in stocking rate maximizes the use of spring pasture and then decreases feed demand throughout the summer in concert with the slump in summer pasture production.**

---

He typically buys bulls in the late fall weighing 950 lbs and sells them the following summer at around 1600 lbs. His average daily gain on these heavy bulls is around 2.75 lbs per day. He could not buy lighter weight cattle because he did not have sheep to de-parasitize his pastures. (Internal parasites in New Zealand have largely become drench-resistant.)

He said that the older bulls grazed at a high grass residual had few problems with internal parasites, however, the low stocking rate meant he had to clip his pastures frequently to keep them vegetative. This necessity for sheep for internal parasite control means that most bull graziers have to access their heavy bulls from the sheep-rich South Island as the North Island has lost almost all of its once large sheep population.

He also fattens 200 or so thin cows he buys opportunistically in local auctions.

"What you need is a way to create a flexible stocking rate to match the seasonal variation in grass growth," he said.

## Organic Beef Production for USA

Another bull grazier I have frequently visited is Phil Taylor of Te Awamutu.

Taylor has a unique "technosystem" development with 300 paddocks that allows bulls in small 20-head herds to be managed as a single large herd. This is done by dividing the ranch into lanes. The bulls' grass consumption is controlled by a single electric wire that runs across all the lanes.

One man can move this one wire. Many small herds are forward moved as one. A similar wire runs across all the lanes behind the cattle to prevent grazing the regrowth and is moved similar to the forward wire.

When the bulls reach the end of their lane, they are walked back to the opposite end of the lane and the forward shifting starts again.

As a risk management diversification, Taylor has been exporting organic grassfed beef to California and has encountered the same learning curve difficulties as American graziers.

He said American organic standards were the most difficult in the world due to their total prohibition of chemical wormers, which are allowed under New Zealand and European organic regulations. As a result, Taylor, who has no sheep on his own farm to naturally de-parasitize his pastures, has had to buy in all of his production from South Island organic custom graziers who still have sheep.

He said the American desire for a year-round fresh supply of marbled meat did not fit the current expertise of New Zealand graziers who are used to seasonal production and no marbling requirements. He said he had been told that they should have to study Argentina's program but they haven't done so yet.

"It's been really tough, but we are determined to figure it all out," he said.

Vaughan Jones, who consults with Taylor and other organic graziers, said that *the herb plantain has shown to help with internal parasites.*

He said New Zealand graziers have their choice of tall plantain for use with cattle or short plantain for sheep.

Keeping soil cobalt and copper high also help reduce internal parasites.

Frequent small amounts of lime help keep the soil earthworm-friendly. Soils rich in earthworms produced far fewer flies.

---

**Few North Americans think about the benefits of fertilizing their pastures occasionally with mined ice cream salt. Soil sodium helps to greatly reduce the incidence of legume bloat and in conjunction with lime helps to soften the grass and prevent plant pulling in new plantings. It also apparently makes the grass taste sweeter and cattle will aggressively seek out salt fertilized grass. French research shows that this salt-sweetened grass makes for an extra flavorful beef.**

Lime also releases molybdenum, which controls bone growth and lime helps prevent footrot.

Surprising to me, Jones said that the most important fertilizer for good animal health was sodium. He said it was critical that soil sodium, potassium and magnesium be in balance to grow white clover well.

He recommends that his graziers replace half of their potassium fertilizer with sodium. He has found that such a 50/50 mix gives the benefits of both.

*Soils that are low in sodium are very prone to bloat.* Velvetgrass is high in sodium and cattle will seek it out when feeling bloated.

Lime and sodium also work together to soften grass and prevent the livestock from pulling out new grass plantings.

He said reactive rock phosphate was now the phosphate of choice for grassland farmers in New Zealand. Acidulated phosphate fertilizers can make cattle sick and do not last long.

The best reactive rock phosphate is Sechura from Peru. The best in North America comes from Wyoming, Jones said.

## The Dinner Knife Steak

As I wanted to taste-drive the very best of New Zealand's grassfed beef, I asked Greg Hefferman — a chef who does beef and lamb cooking demonstrations for New Zealand's Beef and Lamb Marketing Board — for the name of the very best steakhouse in Hamilton and he recommended Tables on the River.

There Vaughan Jones, his wife, Auriel, and I had some really great grassfed Hereford steaks. Interestingly, at New Zealand steakhouses you are not given a serrated edge steak knife as in the USA — only a normal dinner knife. That really shows confidence in the product you are serving.

Tanya Hunt of the New Zealand Beef and Lamb Marketing Bureau told me they have been working for 15 years to improve the tenderness consistency of New Zealand's meat. So far this has been through production criteria and abattoir handling management. Their meat quality assurance primarily rests upon a shear force test of all meat before it enters the retail channels. Any tough meat goes to the grinders.

She said currently one in seven animals was found to be too tough for domestic table meat consumption. This is down from one in four 15 years ago.

## Lessons from New Zealand

■   At the heart of the Kiwi system of grassfed beef production is a seasonal variation in stocking rate that maximizes the use of spring pasture and then decreases feed demand throughout the summer in concert with the slump in summer pasture production.

■   Fall stocking rates are kept low to build stockpiled pasture for winter use.

■   A typical seasonal plan would be to sort off and harvest one-third of the animals in (the North American equivalent of) June, one-third in July and one-third in August. Any stragglers would be finished on brassica-supplemented, autumn pasture before winter.

■   Greater feed efficiency occurs at higher feed intakes because maintenance requirements are satisfied before growth. The greater the feed intake the higher the overall efficiency of gain. Therefore, *it is more grass efficient with heavy cattle in the long term to use a lower stocking rate*, which gives you a higher average daily gain than a higher stocking rate that gives you a lower one.

■   Heavy cattle produce less beef per acre than yearling cattle and so require a price premium per pound to make them competitive in profitability with the alternative of grazing stocker cattle.

■   Annual conversion efficiency is not influenced by winter feeding levels. This is because animals that gain less weight during winter have lower maintenance requirements in spring than heavier animals because they are smaller. This allows for a more rapid spring growth rate.

■ Compensatory gain makes stocker cattle supplementation in winter less cost effective than it appears because one-third of the extra weight gain will be lost by the animal not gaining as well as it otherwise would have gained. However, animals must be at a weight at the start of spring where they will hit their "finish" weight before the worst heat of summer. The New Zealanders accomplish this balancing act by the frequent weighing of cattle.

■ An animal in its final finishing stage should have no restrictions on its input.

## NORTHERN IRELAND

Interestingly, it is not Argentina nor New Zealand but Northern Ireland that has the most pasture-based economy in the world. Despite the bad political publicity, Northern Ireland is incredibly green and beautiful with friendly welcoming people. I have found its graziers to generally be the most technically competent in all of Northern Europe.

Over the years, I have made multiple visits to the Hillsborough Research Station in Northern Ireland near Belfast to study their grass-finished beef production protocol. I found a prototype there that could probably be transferred in whole to winter-wet regions like the Pacific Northwest where winter pugging is a major problem and to deep-snow regions like New England, and Upper Michigan.

Northern Ireland has approximately a seven-month grazing season that runs from early May through mid-November. This green season is coupled with a five-month period of feeding stored forages off pasture on concrete feed pads. Cattle are available for slaughter year around for domestic use from this production system. English breed cattle for domestic consumption are typically 18 to 24 months old at harvest.

This winter feeding is not due to the cold weather as temperatures seldom fall below 20°F but because of the wet. The soils in Northern Ireland are over 10% organic matter and will not hold up an animal as heavy as beef cattle during the winter rains.

Their base pasture is primarily perennial ryegrass. The most common stored forage is grass silage harvested from excess spring pasture. Very little hay is made in Northern Ireland due to the frequent showers. Most silage making is done by professional contractors. Young cattle going back to grass in the spring are typically fed a diet of nothing but grass silage in the winter. *Cattle scheduled to be harvested in the winter are fed small amounts of beet or citrus pulp to increase daily weight gains to the 1.8 lbs necessary for finishing.*

Unlike Argentina, Northern Ireland uses almost no supplemental annual crops for summer. This is because the cool summer temperatures keep the perennial grass digestible in the summer. Summer temperatures seldom rise above 75°F and plant lignification is not a problem.

Research at Hillsborough has shown that pastures with good white clover stands plus 50 lbs of supplemental nitrogen can produce within 80% as much grass as 300 lbs of applied nitrogen. However, *it is very difficult to keep clover in their pastures under their management regime of setting aside the vast majority of their farm for silage cutting every spring.* This is because the clover gets shaded out by the tall grass.

The Irish typically do not use rotational grazing with beef cattle as they do with their dairy cattle. Ryegrass is not harmed by continuous grazing in their moist, cool summers and so continuous grazing is heavily used in Northern Ireland. In contrast, in summer-hot Australia the grazing of perennial ryegrass

---

**Vegetable oils and meals, tallow from grassfed animals, molasses and sugar beet pulp could all be used as energy supplements with no negative effect on CLA concentrations in the meat and milk of the animal.**

**Dr. Tilak Dhiman**

must cease altogether when daytime temperatures climb over 90°F to avoid stand thinning.

However, the Irish do not just set stock and forget. They use a management-intensive system called "buffer grazing" whereby a one-wire, temporary electric fence is moved to expand or contract the size of the pasture to meet the animals' feed demands. This "buffer" comes from ground set aside for harvest as pasture silage.

*Their goal is to keep the grass at a nearly constant height between four and five inches for the entire green season.* During periods of grass shortage, an additional area, which was previously designated for silage must be added to the grazed portion to maintain the desired grass residual.

In contrast during times of extremely rapid grass growth, the size of the grazed area is made much smaller and more of the area is harvested for silage. As a general rule only 25% of the total pasture will be used only for grazing and 75% will be cut for silage at least twice.

This gives the grazier a flexible stocking rate that is high in the spring and much lower in the fall. Such a flexible stocking rate is necessary not only due to grass growth rate changes but due to the fact that as cattle get heavier they require more grass.

The Irish estimate that the fall stocking rate with finishing cattle should be only 25 to 40% of the spring one.

## Grass Silage

For finising weight animals, the most important factor in the feeding quality of grass silage is the stage of maturity at which the crop is harvested.

Cutting the grass prior to maturity can produce an excellent feedstuff capable of producing gains of a pound and a half with no supplementation and over two pounds with a small amount of supplementation. A three-cut system with a late May, early July and late August harvest produced the highest quality grass silage and the same total tonnage as a two cut, early June and late August harvest.

The Irish figure that between 3.5 and four beeves can be fed for 150 days per acre of ryegrass silage.

A high quality silage is particularly important if winter finishing is desired. In Hillsborough research, three-cut silage supplemented with five pounds of concentrate produced an average daily gain of 2.6 lbs per day versus 2 lbs a day for the two-cut silage.

*The Irish figure that their grass silage-minimum-concentrate system produces finished beef at about 60% more net profit than a high-concentrate-minimum forage, USA-style finishing system does.*

The Irish have found that a low-cost flail-type silage harvester and a direct-cut produces a high quality silage more consistently than a precision-cut pick up unit harvesting pre-mown and wilted grass.

While the wilting and precision-cut has resulted in a higher daily intake by the cattle, this was not reflected in increased animal gain due to the lower quality of the wilted silage.

In fact, due to frequent showers during the wilting process, the wilted silage frequently produces animal performance 10-12% below the unwilted, direct-cut silage.

Their research found that the optimum level of performance for steers weighing 550 to 750 lbs that are to be returned to pasture is 1.5 lbs per day. Steers overwintered on grass silage alone and harvested the following winter after a summer of grazing weighed two pounds more than those supplemented through the winter with three pounds of concentrates per day.

The desired 1.5 lbs per day level of gain can be produced by high quality grass silage with no added supplements. On the other hand, for lightweight stocker cattle weighing 250 to 450 lbs, supplementing high quality grass silage with 3.5 lbs of an 18% protein supplement was found to significantly increase animal performance during the feeding period without hurting subsequent spring gains on pasture.

**Finishing Cattle on Silage**

The amount of supplementation for silage finished cattle depends upon the breed, sex and degree of fatness desired. Medium frame, easy-fattening, English-type cattle show a poor growth response to over five pounds per day of energy supplement (normally beet or citrus pulp).

For finishing heavy 1000 lb plus cattle, no increase in animal performance was found by increasing the protein content of a supplement to over 10%. The grass silage alone can provide all the protein these heavy beeves need. What is needed is a carbohydrate. Extreme care must be used to not feed enough carbohydrate to cause a rumen shift and thereby decrease the digestion of the forage.

Consequently, *the feeding value of molasses and all carbohydrates declines with the amount fed in a high forage diet.* The Irish have found that around 5.5 lbs per head per day is the maximum level for 1000 lb-plus beeves.

Research at Hillsborough found no benefit to mixing the supplement with the silage (TMR) versus feeding to two feeds separately as long as the amount of supplement being fed was less than 6.5 lbs per day.

This chart shows that non-starch feeds can provide as much energy as grains.

| Feedstuff | Protein content (%) | Energy content (MJ/kg) |
|---|---|---|
| Corn | 8 | 12.2 |
| Dry Barley | 9.5 | 11.1 |
| Corn gluten | 18 | 10.9* |
| Citrus pulp | 6 | 10.9* |
| Molasses | 4 | 8.0* |
| Molassed Beet Pulp | 8 | 10.6* |
| Extruded Soybean meal | 44 | 11.6* |
| Whole cottonseed | 32 | 10.4* |

* Non starch feeds

**An Irish Prototype in Maine**

Roger and Linda Fortin's Little Alaska Farm in Wales, Maine, is a good example of the Irish prototype in North America.

Their production program features a May until mid-October direct-graze pasture program combined with winter-fed grass silage. *This prototype has allowed them to supply a Massachusetts abattoir with four forage-finished beeves a week for over four years.*

Pastures are a mix of orchardgrass, timothy and volunteer white clover.

The Fortins have a 100-cow herd of mostly Red Devon (also called Ruby Reds) breeding and also buy in Devon-bred yearlings from neighboring farms. Roger likes the Red Devon for its easy marbling on grass.

The beeves are currently harvested at 20 to 28 months of age with an average carcass weight of 600 to 650 lbs and sold to Hardwick Beef in Hardwick, Massachusetts, for a carcass weight of $1.75 a pound.

The wilted grass silage must be supplemented with six pounds of citrus pulp per head per day to get the average daily gains above the 1.8 lbs per day needed for marbling.

Such consistent high gains are important as Hardwick Beef requires all of its grassfed beef to be in the high Select to low Choice grade.

As a non-starch energy source, citrus pulp meets Hardwick Beef's grassfed protocol and is locally available as a dairy feed. Roger said his farm used to be a dairy and so was all set up for silage storage and feeding.

**Direct Cut Silage Preserves Valuable Fatty Acids**

Research in Wales found that direct-cut, unwilted silage preserves much more of the healthful Omega-3 fatty acids than hay or wilted silage. As a result, this system is gaining in popularity with grass-finished beef producers.

Direct cutting uses a flail-type chopper unit to cut the

grass at relatively long lengths. This long length is better than fine chopping as it allows the animal to regurgitate the silage for rumination. Also, the less damage there is to the grass the less nutrient loss there is from effluent weeping.

Fine chopping is only necessary when upright storage structures are used. With silage stacks, *the longer the cut the better.*

High rpm flail choppers, such as the Alpha-Ag Lacerator, blow the cut grass into a following wagon or a truck alongside without the use of augers, which can damage the forage.

Approximately one tractor horsepower is required for each width of cutting face with this unit. For example, a six-foot-wide unit (72 inches) would require at least a 72-horsepower tractor.

Direct-cut silage is a one-pass operation whereas wilted silage requires at least two passes and two separate pieces of machinery — a mower and a chopper. This saves both time and fuel and requires less machinery investment.

For maximum quality, grass silage should be cut at the exact same stage of growth and height where it would be grazed for a high average daily gain. Too much grass silage is cut when it is too mature for high animal performance to maximize tonnage.

Silage cut in the afternoon is higher in sugar content than that cut in the morning and will produce better animal gains.

---

**In using vegetable oils and meals, flax oil, ground flax meal and canola oil and meal are the best to use as they are very high in Omega-3 fatty acids. A European research study grazing steers on ryegrass supplemented with ground flax meal had higher Omega-3 levels than un-supplemented ones.**

**Jo Robinson**

While some say silage can be cut in the rain, this is not advised. A light shower in the middle of a cutting may not hurt but do not try to cut silage in a pouring rain storm.

The cut grass is unloaded at the site where the silage stack will be built. No side walls or hard-surfaced floor is needed with the vacuum silages. There is no need for tractor packing to exclude air. It can be built directly upon the ground.

### Determining the Size of Silage Stack

The size of the silage stack will be determined by the size of the plastic sheeting used.

There are plastic sheets available that are as large as 50 feet by 200 feet and as small as 24 feet by 100 feet. However, the 40 x 100 size is probably the most popular. This is the equivalent of 40 to 50 large 1500-round bales and can easily be made in a day.

Making and sealing a stack on the same day minimizes wilting and subsequent Omega-3 loss.

The Europeans believe that multiple small stacks also lessen the risk of silage loss from an accidental plastic puncture.

Some have found shaping the pile with a front-end loader useful in making a smooth surface that will prevent the holding of rainwater puddles on the stack that turns to ice in winter. However, running over the cut material is not only unnecessary but will cause damage to the grass material and promote weepage.

The best site for a silage stack is where there is a slight incline for drainage. Also, building the stack close to where it will be harvested saves time and fuel.

### How to Make Vacuum Silage

A perforated pipe is centered in the site area. When the stack is about half done, lay a 10-inch piece of 4-inch sewer pipe on the ground so that the pipe extends from the center of the stack to the edge. Unload or push the next load of silage on top of the pipe. The stack is then continued until completed.

White plastic is then pulled over the stack. White plastic is preferred to keep the silage stack cooler. ***Black plastic can result in extremely hot temperatures that can caramelize the plant sugars and lower its feed value.***

Sand, dirt, ag lime or sandbags are then placed around the edges of the plastic to help create a seal. A small vacuum pump is connected to the pipe, which was laid in the stack earlier and the air is removed from the silage stack.

After a few minutes of vacuuming, the pump is turned off and the extraction pipe is sealed. The silage is then allowed to ferment until it is used. Unlike hay, which will lose quality over time because it is always exposed to the air, a vacuumed stack that stays airtight can be stored indefinitely without nutrient loss.

In New Zealand, sealed silage stacks intended for long-term storage are covered with a foot or more of dirt, and grass is planted on them. This helps prevent animal and rodent damage to the stack's plastic covering.

**Feeding Grass Silage**
Once the silage stack is opened the entire face should be removed each day to prevent spoilage. However, the stack does not have to be resealed as long as the use is continuous.

---

**Using flax or canola meal with grass silage not only provides energy but helps replace the Omega-3 fatty acids that are often lost in stored forages due to wilting.**

**While soybean oil and meal are relatively high in Omega-3, cottonseed oil and meal are very high in Omega-6 fatty acids and should not be used. Most cottonseed products also contain gossypol that lowers male ruminants' fertility.**

**Jo Robinson**

A silage grab or slice will help prevent deterioration as it is designed to leave a smooth silage face with long-cut material.

A New Zealand-designed, multi-purpose, transport-feedout wagon is available from Alpha-Ag in Illinois that is designed specifically to feed out grass silage.

*An alternative is to stretch an electric wire across the silage face and let the animals feed directly.* Solid fiberglass posts are driven horizontally into the silage face. The rate of silage feeding can be controlled by periodically tapping the posts deeper into the silage face.

If direct-feeding is to be used, the silage stack should not be built taller than four feet.

If self-feeding is used in winter, it is best to build the stack on a concrete slab, otherwise it will become quite boggy at the silage face. This is particularly true during the early spring freeze-thaw period.

In areas where the ground freezes in winter, direct feeding at the silage face works well when the ground is frozen solid. No problems with the stack face freezing solid have been reported with such continuous direct-feeding although some have reported a "crunchy" texture.

Orienting the stack in a North/West, South/East direction and feeding from the southeast end will head off potential problems. Adding a little salt to the cut material will also help with face freezing.

In Argentina, direct face feeding is most often used in the late summer and early fall while waiting for winter annuals to be

---

**Sugar beet pulp has the added advantage of being very high in anti-oxidants. Anti-oxidants are what give grassfed beef its long shelf life and bright red color. A recent nutrition study found that beet is one of the top five vegetables for anti-oxidants.**

**Jo Robinson**

ready for grazing. Under such dry-season conditions, there is very little problem with bogging at the face.

One problem with direct face feeding is that it concentrates the animals' manure in a small area. To get a more even distribution of manure, feeding the silage from a feedout wagon on the pasture under an electric wire has largely replaced direct feeding in most of the world.

Pueblo, Colorado, grass finisher, Russ Maytag, direct feeds his grass silage. He said his background was in stocker cattle and it took a little mental tweaking to get out of the "rough them through the winter and pop them out on mountain pasture" mindset.

"Now, I want my animals gaining well year around so they will be tender."

To help do this he has replaced winter hay feeding with direct-cut, unwilted grass silage. This is used to supplement, direct-grazed, stockpiled, irrigated fescue and orchardgrass, which is his primary winter feed.

His home ranch in dry, southeastern Colorado lies at the same latitude as Virginia and Kentucky and has very little winter snow. Unfortunately, this latitude also produces very hot summers with daytime temperatures of 100°F or higher.

For this reason, his steers are taken to mountain pastures as soon as they green up in the spring. The lowland pastures are then cut for grass silage.

High summer temperatures have made it necessary for him to "tweak" the recommendations for making un-wilted grass silage.

"At 100°F, you have to get the air out of the silage very quickly to prevent over-heating and losing your silage quality. (The plant sugars will caramelize at high temperatures.)

"While the recommendations are to not pack the silage stack but just suck the air out with a vacuum pump, we found a slight packing really helps to maintain silage quality if it is made under hot weather conditions. This is particularly true if the stack is going to be made over two days rather than one."

Maytag uses a small Bobcat with a front scoop to shape the stack. He then runs over it several times to help push the air out. "We don't pack it to the extent you would if you were making un-vacuumed silage. It's just a little packing."

The silage is self-fed in the winter.

## Self Feeding Silage

To do this, a strip of poly-tape is stretched across one of the faces of the silage stack with the fiberglass fence posts driven into the silage face so that they are horizontal to the ground. The posts are driven in far enough that the cattle can reach the silage face.

Every day or two, Maytag taps on the end of the fence posts to move the tape closer to the silage face. This allows the cattle to eat their way through the pile and self-feed themselves. There is no need for a tractor or feed-out wagon.

In the dry Colorado climate mud at the face is not a problem.

Early in the winter the cattle are given access to both the stockpiled pasture and the silage. He said it takes from ten days to two weeks for the cattle to get adjusted to the silage.

"At first they will eat just a few nibbles and return to the pasture but once they get used to it they just love it and will lick it off the ground."

He did not have any weight figures on what the animals' average daily gain was from the silage but said it was better than alfalfa hay.

"It's a great system but it takes some tweaking in hot weather areas," he said.

David and Lynda McCartney of Coleman, Michigan, have used direct cut vacuum silage for several years. In 1998 they were awarded a NCR SARE grant to study the relative benefits and costs of alternative forage harvest systems.

The findings from the McCartney study were that direct cut vacuum silage had these advantages over conventional methods of harvesting forage:

- 70% less harvest time.
- Less weather risk.
- Less capital investment.
- Less downtime.
- More profit.

McCartney's cost study summary found these benefits of direct cut silage:

- **Capital costs were 1/3 of conventional silage equipment and about ½ as much as equipment for making hay.**
- **Storage costs were slightly less than baleage but more than dry hay in round bales.**
- **Operational costs were about 1/3 the cost or less of any other method studied.**
- **Total costs per ton of dry matter harvested can be 1/3 the cost of making dry hay.**

## Mud, Not Snow, Is The Biggest Winter Problem

Winter mud is actually a far bigger problem than deep snow. In deep snow areas, buying a snow plow may be far more cost effective than making and feeding a stored forage.

John Cross runs the 13,000 acre A-7 Ranch near Nanton, Alberta, south of Calgary. Surprisingly, he neither makes nor feeds any hay on his ranch and relies completely on direct-grazing for his feed supply year around.

During severe winters deep snow cover can extend for as much as 90 days at a stretch. He has learned that it is far more cost-effective to move the snow off the feed than to feed hay on the snow.

He stockpiles his grass and rents a neighbor's tractor to blade the snow off the grass during those infrequent heavy snow cover days. The pastures are pushed clear of snow in 30-acre increments. This takes two to three hours a day or roughly about the same time feeding hay required.

By using temporary fence moved frequently, Cross can get 70 to 90% utilization of the grass. All of the manure is naturally recycled back onto the pasture.

John owns the blade, which must have rounded corners to prevent pasture damage, but has sold the rest of his equipment. The 800-cow ranch is operated by himself plus one employee and his daughter.

A key element in making this low-labor, minimal machinery successful was a switch to July calving. He said this has made the whole ranch much simpler to operate.

Calves are now over-wintered on the teat with their mothers. This is much easier on the calves but a little tougher on the cows.

The ranch is currently subdivided into 50 permanent subdivisions. Stockwater is provided by portable stock tanks on 70% of the ranch. These wheeled tanks shift with the cattle.

Cross started producing grass-finished beef on a small scale about three years ago.

While the green season in southern Alberta is only three to four months long, he can get three pounds a day gain for nearly the whole season. His pastures are a mixture of fescue, timothy and brome.

During the winter on the stockpiled pastures, he can get two pounds a day with four to six pounds of non-starch energy supplement.

### Keeping Your Cattle Out of the Mud

There are alternatives to an Irish-style concrete feeding pad in wet weather regions.

While most of the Pacific Northwest west of the Cascades can grow excellent winter pasture, and in particular winter annual pasture, grazing these high quality pastures is a major problem due to the extremely wet winters that produce severe pasture pugging problems.

Winter mud is also a problem in Northern California due to the widespread lack of sod-forming grasses. Most ranges in California are wide-spaced cool-season annuals and offer little hoof support.

Research in New Zealand has found that severely pugging

soils in the winter will decrease spring and summer production by as much as 30% and can actually lower pasture productivity for years.

There are but two ways to deal with mud.

One is to *get the water off the pasture as quickly as possible with proper drainage.*

This can be through below ground tiling, or through the creation of surface ditches.

In New Zealand on very wet natured soils they use a method called "hump and hollow" whereby soil is dug from one area to provide drainage and the removed soil is piled nearby to create a well drained area for the cattle to rest upon.

And two, is to *get the animals off wet pasture after allowing them a short graze.* In areas, where mud is only an occasional problem this is the cheapest solution.

### Short Grazing Periods

Three hours of grazing on stockpiled cool-season grass is all dry cows need to get all the grass they need.

Research in Argentina has found that steers being finished on winter annuals will not show a decrease in average daily gain if fed half their daily dry matter in alfalfa hay off the pasture. A finishing weight steer consumes about 3% of his bodyweight per day in dry matter.

Grass hay should never be used with finishing cattle.

It is important that the finishing steer have daily access to the winter annuals or his gains will fall below those needed for finishing (1.8 lbs per day).

### Have a Sacrifice Area

*Mounds of sawdust, wood chips, bark peelings or agricultural lime can provide for resting areas when the cattle are "off" the pasture. These are called "stand-off" areas.*

However, such areas should not be used for long-term hay or silage feeding but only for standing off the pasture during short-term rain events.

For long-term feeding, a concrete feed pad covered with chips or sawdust should be used. This will facilitate the capture of the manure for composting and reuse.

Cattle whose hair becomes caked with mud and manure require much more energy to stay warm.

### Get the Cattle off the Pasture During the Rain Event

The majority of pasture damage occurs during the actual rain event. This is when it is most important to have the cattle off the pasture.

Also, never leave cattle on a wet pasture overnight.

### Use a Backfence

Many winter graziers do not use a backfence due to the lack of regrowth in the grazed pasture. However, this allows these short-grazed pastures to get pugged by animals walking on them.

### Never Have Hungry Cows

A hungry cow will eat up to 90% of her daily feed requirement in two to three hours. Once you see animals stop grazing and lay down to ruminate, get them off the pasture!

*Always feed supplemental hay after they graze. Not before.*

### Budget out Large Paddocks

It is important that you budget out the grass so that the cattle do not get used to lax grazing in winter and slow their rate of consumption. Remember, we want to promote aggressive grazing in wet weather.

### Walk the Cows over Long Grass

Pastures with long grass can withstand pugging far better than those with short grass. When budgeting out a large paddock for a three-hour graze, always walk the animals over the front of the paddock and start your grazing at the back.

Too many graziers will start from the gate and work toward the back. This means the cows will be walking over short, easily pugged grass.

## Use Argentine Style Gateless Fences

The worst mud occurs at the gate to a paddock. It is important that the cattle do not drag mud onto the grass from their feet as this ruins it for grazing.

Argentine style one-wire fence whereby the wire can be lifted with a pole are far preferable to permanent gates in wet-weather country.

This allows you to vary where the cattle enter the paddock.

If the Vela pole is to be left standing, it must be energized or the cattle will knock it down. If it is only used for a quick paddock shift a length of PVC pipe can be used.

## Re-grass Immediately

On areas that become severely pugged, new grass and legume seed should be added quickly. Annual ryegrass and birdsfoot trefoil are two forages that grow exceptionally well on bared mineral soil.

## Tips from Ireland

■ Winter feeding is not due to the cold in Ireland, but to wet conditions. As a result, Irish graziers have become masters with silage making.

■ Young cattle going back to grass in the spring are typically fed a diet of nothing but grass silage in the winter. Cattle scheduled to be harvested in the winter are fed small amounts of beet or citrus pulp to increase daily weight gains to the 1.8 lbs necessary for finishing.

■ Cutting the grass prior to maturity can produce an excellent feedstuff capable of producing gains of a pound and a half with no supplementation and over two pounds with a small amount of supplementation. A three-cut system with a late May, early July and late August harvest produced the highest quality grass silage and the same total tonnage as a two cut, early June and late August harvest.

■      For finising weight animals, the most important factor in the feeding quality of grass silage is the stage of maturity at which the crop is harvested.

■      The long length of grass in silage making is better than fine chopping as it allows the animal to regurgitate the silage for rumination.

■      For maximum quality, grass silage should be cut at the exact same stage of growth and height where it would be grazed for a high average daily gain.

■      Silage cut in the afternoon is higher in sugar content than that cut in the morning and will produce a better animal.

■      The best site for a silage stack is where there is a slight incline for drainage. Also, building the stack close to where it will be harvested saves time and fuel.

■      Once the silage stack has been opened the entire face should be removed each day to prevent spoilage.

■      If self-feeding is used in winter, it is best to build the stack on a concrete slab otherwise it will become quite boggy at the silage face. This is particularly true during the early spring freeze-thaw period.

■      Orienting the stack in a North/West, South/East direction and feeding from the southeast end will head off potential problems. Adding a little salt to the cut material will also help with face freezing.

# CHAPTER 8

# Start with Heifers

I recommend that new grass finishers start with heifers because they are easier to finish than steers. If you are dead set against using annual forages and don't want to overwinter heavy yearlings for their second winter, you may want to consider an all-heifer program.

Compared to steers, heifers mature earlier and marble easier. As a finishing target guide, figure a heifer's mature weight as 80% of a steer brother's weight. However, heifers fed on high-quality grass from weaning can reach fat cover and marbling targets as early as 70% of the steer slaughter weight.

Therefore, if a steer needs to be 1100 lbs at slaughter to marble, a heifer of similar genetics would reach the fattening target for harvest at 800 to 900 lbs. This 200-lb difference in target weight between heifers and steers could be a great problem solver in your grass-finishing program as a 200-lb difference means you need 100 days less of quality feed. This is a critical point due to the short finishing period of perennial forages.

Keep in mind, *when I say perennial forages I mean a legume-based forage. In other words, a perennial grass with a perennial legume like alfalfa or white clover.*

The Table on ADG of Heifers and Steers is from Anibal Pordomingo and shows the liveweight differential of heifers and steers under a similar weight gain expectation. These are for medium-frame English breed cattle.

For comparison, if both sexes enter the growing-finishing program in October with similar weights (steer calves are a little heavier) after weaning, and follow a similar rate of gain as suggested in the Table (average daily gain = ADG), it will take 8 months to finish the heifers at above 800 lb liveweight, and about 12 months to finish the steers on the 1000- to 1100-lb range if there are no quality or quantity feed restrictions.

Steers may even take longer if fall gains are lower than those established in this table or if the animals are of a larger phenotype. At the gains shown, the steers would not be truly finished due to the low summer and fall average daily gains. On perennial-only pastures, North American steers will end up not

**ADG of Heifers and Steers Table Simulated weight evolution and gain (ADG) of heifers and steers under a similar daily gain scheme.**

| Month | ADG lb/day | Heifers | Steers |
|-------|------------|---------|--------|
| | | Live weight, lb | |
| October | - | 450 | 480 |
| November | 1.0 | 481 | 511 |
| December | 1.0 | 511 | 541 |
| January | 1.0 | 542 | 572 |
| February | 1.5 | 589 | 619 |
| March | 1.5 | 634 | 664 |
| April | 2.0 | 696 | 726 |
| May | 2.0 | 756 | 786 |
| June | 2.0 | 818 | 848 |
| July | 1.5 | - | 894 |
| August | 1.5 | - | 936 |
| September | 1.5 | - | 983 |
| October | 1.5 | - | 1029 |

Source: Dr. Anibal Pordomingo

being truly finished until the following spring when they are two years old.

Selling finished heifers in late spring and early summer gives the opportunity of making a good use of spring grass and de-stocking early so that you can stockpile more grass for winter for your cows and calves. However, grazing un-spayed, early maturing heifers can be a management headache.

Good quality grass makes intact heifers sexually preco-cious and they will spend a lot of time riding one another. Of course, they can't graze and do this so intake declines, which affects weight gain and marbling rates. Therefore, *fattening heifers that cycle regularly could take 60 days longer than spayed heifers.*

If at all possible, a perennial pasture, heifer-based operation should plan on selling as many fat heifers as possible at the end of spring and the first half of summer.

If you are unfamiliar with the finishing process, it would be wise to chart the growth curve on paper as they do in Argentina. You will find a forage average daily gain chart in the chapter on The Forage Chain.

Anibal said the most relevant information needed is: a) the forage the heifers will be on from weaning to harvest, and b) the expected weight gains, at least on a monthly basis.

*We need to have an estimate of weaning weight, a desired final liveweight, and a target harvest month.*

The forage used will tell us about possible weight gains, which will in turn define the length of the growing-finishing process and the seasonal feasibility of producing fat-based gains during the last 60 to 90 days before harvest.

However, the harvest month is very important too.

A fair forage can take the heifers to the slaughter weight, but it could be too early or too late in the year. There-fore, we need to equate the month(s) too.

This may lead us to review: a) the forage (to improve gains), b) the weight and time at weaning (or purchase) in the short term, and c) genetics in the longer term.

As an exercise, let us say we plan on a weaning weight of 450 lbs, and a final weight of 800 lbs to happen in June.

If the forage chain gives as an average daily gain of 1.5 lb/day, then we could reach the 800-lbs weight in 8 months.

If we wean in October with 450 lbs, the heifers will have reached the targets in June (ADG Table). And, if at least the last 60 days produce near 2 lbs/day, most likely the gain will have reached the minimum fat cover and marbling needed.

***Weaning weight plays a huge role in the success of grass finishing.***

The ADG Table shows the evolution of liveweight of heifers under a similar growing-finishing program, but a different initial (weaning) weight. The lightest one at weaning reaches the target in 8 months. But, if the heifer was 100 lbs heavier (the heaviest in the ADG Table) she would reach the final weight a month and a half earlier.

As you work through such exercises, you will see that using only a perennial forage is hugely limiting and greatly increases the risk that some heifers — and most steers — will not finish at all until the following spring.

Keep in mind in an all-heifer program, adding only a summer annual would allow you to finish any tail-end heifers before winter and some steers. Adding a winter annual would allow the heifers to arrive at a heavier weight in the spring thus insuring they would finish by the end of June on the perennials.

---

**Mac Magruder of Potter Valley, California, didn't have a problem coming up with an extended harvest from one calving season due to using a mix of heifers and steers.**

**"We harvest animals between 18 and 26 months of age. Nothing is younger than 18. We spay all of our harvest heifers, which makes them finish faster. This allows us to harvest heifers in the spring and steers in the summer.**

Some could even be harvested earlier.

In most instances in a heifer-only program, if you have to choose, adding winter annuals will give you the most advantages.

# CHAPTER 9

# The Forage Chain

While farmers' markets, internet and mail order markets can make do with a seasonal, frozen product, most "gourmet" customers want a fresh, unfrozen meat. Keep in mind, that for over 100 years American consumers have had access to chilled fresh beef on a year around basis. There is no collective cultural memory of beef being a seasonal product in the USA.

Two San Francisco Bay Area "gourmet" restauranteurs who have pioneered putting grassfed beef on their menus told me they like everything about California grassfed beef except its current seasonality.

"Availability is the big hurdle grassfed producers will have to clear," said Thom Fox, head chef of the Acme Chop House in San Francisco.

"It just absolutely must be available more times in the year. It doesn't have to all come from the same producer, but it must come more often to have any real impact on the consumer."

Due to California's summer-dry, Mediterranean climate, most producers without irrigation only have well-finished grassfed beef for a couple of months in the late spring.

"We have made a commitment at Chez Panisse to use nothing but grassfed ruminant products," explained Sue Moore of Alice Waters' Chez Panisse restaurant in Berkeley. Sue is always in search of gourmet quality grassfed beef producers.

"However, this decision has meant that beef is off our menu for much of the year and our customers aren't happy about it."

She said they currently serve grassfed beef, lamb and bison produced in Northern California. "Local-ness" and "freshness" of the food served is a big part of Chez Panisse's cachet.

*Moore said that while customers were accepting of the idea of lamb and bison seasonality, they were not so accepting about beef seasonality.*

"People just expect beef to be available year around," she said.

This can be done with a forage chain.

On the next two pages you will find a chart of most of the popular temperate forages and the average daily gain each is capable of producing with finishing weight cattle (1000 lbs).

This chart was assembled by Anibal Pordomingo and represents the mid-latitude region of Argentina that has a climate and rainfall similar to east-central Oklahoma and Kansas. The months have been changed to their Northern Hemisphere equivalent.

It's important for you to study this chart carefully.

Note that *most cool-season grasses with a legume can produce finishing gains in the March to June period.*

Also note the clear gain superiority of white clover to red clover in summer average daily gain, of perennial ryegrass to orchard and fescue, the excellent winter-season finishing gain of winter annuals, the high-summer gains of Eastern gamagrass and the total lack of finishing gains from any listed forage in October and November.

Anibal said that if you absolutely must have finished cattle in those two low-forage-carbohydrate months, the only forage capable of producing them would be late-planted green leaf corn.

Of course, before you try to chart where you want to go it is important to know where you are first.

| | SUMMER | | | FALL | | | WINTER | | | SPRING | | |
|---|---|---|---|---|---|---|---|---|---|---|---|---|
| | J | J | A | S | O | N | D | J | F | M | A | M |
| **PERENNIAL PASTURES** | | | | | | | | | | | | |
| Alfalfa + orchardgrass | 2.2 | 1.5 | 1.8 | 2.0 | 1.5 | | | | | | 2.0 | 2.4 |
| Alfalfa + fescue | 2.0 | 1.8 | 1.5 | 2.0 | 1.5 | 1.5 | | | | 2.0 | 2.2 | 2.0 |
| Alfalfa + western wheatgrass | 2.0 | 1.5 | 1.5 | 1.8 | 1.2 | 1.4 | | | | 2.0 | 2.2 | 2.2 |
| White clover + ryegrass | 2.4 | 2.0 | 2.2 | 2.0 | 1.5 | 1.5 | | | | 2.0 | 2.4 | 2.4 |
| White clover + orchardgrass | 2.0 | 1.8 | 2.0 | 1.5 | 1.5 | 1.5 | | | | 2.0 | 2.4 | 2.3 |
| Red clover + peren. ryegrass | 1.0 | 1.8 | 1.8 | 1.8 | 1.6 | 1.5 | | | | 2.0 | 2.2 | 2.3 |
| Red clover + fescue | 1.5 | 1.5 | 1.5 | 1.8 | 1.5 | 1.4 | | | 1.5 | 2.0 | 1.8 | 1.5 |
| Red clover + orchardgrass | 2.2 | 2.0 | 1.6 | 1.8 | 1.6 | 1.3 | | | | 2.0 | 2.4 | 2.4 |
| Gammagrass | 2.4 | 2.0 | 2.4 | 2.0 | 1.2 | | | | | | 2.8 | 2.8 |
| Johnsongrass | 1.5 | 1.0 | 1.2 | 1.2 | 1.0 | 0.5 | | | | 1.5 | 1.8 | 1.8 |
| **ANNUAL WINTER PASTURES (grasses)** | | | | | | | | | | | | |
| Cereal rye | 2.0 | 1.8 | | | | 1.5 | 2.0 | 2.4 | 2.0 | 1.8 | 1.5 | |
| Annual ryegrass | | | | | | | | 2.0 | 2.4 | 2.5 | 2.5 | 2.2 |
| Wheat | | | | | | 1.5 | 2.0 | 2.0 | 2.5 | 2.2 | 1.8 | |
| Triticale | | | | | | | 2.0 | 2.2 | 2.5 | 2.5 | 2.0 | 1.5 |
| Cereal oats | | | | | | 1.8 | 2.2 | | 2.0 | 2.4 | 2.0 | 1.6 |
| Barley | | | | | | 1.6 | 2.2 | 2.2 | 2.0 | 2.4 | 1.8 | |

| | SUMMER | | | FALL | | | WINTER | | | SPRING | | |
|---|---|---|---|---|---|---|---|---|---|---|---|---|
| | J | J | A | S | O | N | D | J | F | M | A | M |
| **ANNUAL WINTER LEGUME + GRASS** | | | | | | | | | | | | |
| Wheat + vetch | | | | | | 1.5 | 2.0 | 2.2 | 2.2 | 2.4 | 2.2 | 2.0 |
| Oats + vetch | | | | | | | 1.8 | 2.0 | 2.4 | 2.4 | 2.2 | 2.0 |
| Triticale + yellow trefoil | | | | | | 1.5 | 2.0 | 2.2 | 2.3 | 2.5 | 2.2 | 2.0 |
| Cereal Rye + trefoil (white) | | | | | | 1.5 | 2.0 | 2.5 | 2.2 | 2.0 | 1.8 | 1.8 |
| **ANNUAL SUMMER PASTURES** | | | | | | | | | | | | |
| Sorghum sudan | 1.8 | 2.0 | 2.0 | 1.8 | | | | | | | 1.8 | |
| Corn plant | 2.2 | 2.5 | 2.8 | 2.8 | | | | | | | | |
| Pearl millet | 1.5 | 1.8 | 1.5 | 1.0 | | | | | | | | |
| Soybeans | 2.2 | 2.0 | 2.2 | | | | | | | | | 2.0 |

## The Forage Chain Chart

Expected liveweight gains (lb/day) for grassfed, medium-frame beef steers in temperate sub-humid environments. Data was compiled from research trials and commercial operations throughout central Argentina. Rates of gain reported assume a normal grown, non-nutrient-restricted plant. Months represent those of the Northern Hemisphere.

Dr. Anibal Pordomingo

If your land is coming from long-term cropping, it is unlikely to have enough soil nitrogen in it to successfully grow high carbohydrate annual crops. In that case, your forage chain will start with an extended period of building soil organic matter with a high legume percentage perennial forage. This soil building period may require as long as five years.

If you plan to farm organically it is critical that you use this perennial pasture period to get any severe weed problems under control. You can do this with a combination of mineral fertilization, grazing management and strategic mowing.

The two legumes of choice are (1.) white clover or (2.) alfalfa if your rainfall is less than 40 inches per year.

Most American graziers are afraid of such high-legume-percentage pastures, however some Americans have been using them for many years.

## Alfalfa Grazing Pioneers

The Abbey of the Holy Trinity is located high in the Utah mountains near Ogden. This Abbey of 24 Catholic Brothers supports itself by making and selling mixed grass/alfalfa hay, beef and honey.

I visited Brother Nicholas there who supervised the Abbey's pioneer alfalfa grazing program.

He said that rotational grazing and the grazing of alfalfa was being promoted by Utah State in the 1950s with few takers. However, direct grazing sounded like a good idea to the Brothers who were struggling to survive selling $15 a ton grass/alfalfa hay.

Following Utah State's advice they subdivided a hay field into 10-acre paddocks with electric fence. A paddock rotation was designed to give the alfalfa 35 to 40 days to recover following grazing.

He said the Abbey liked an orchardgrass, alfalfa mix.

Fall-weaned calves were wintered on alfalfa hay and started grazing when the orchardgrass in the hay fields was six inches high the following spring. Grazing traditionally started on

May 1. By that time the fall-weaned Hereford calves were yearlings and weighed around 700 lbs.

The yearlings would be allowed to have access to free-choice alfalfa hay while they were becoming accustomed to grazing the low dry matter spring pastures. Because the spring pastures were predominantly grass, he said *bloat was not a great problem if a gradual transition from hay to grass was followed.*

As the weather warmed, the pastures grew in legume percentage until the alfalfa became dominant in mid to late summer. He said that was the time you needed to be watching your cattle closely for bloat.

Average daily gains on the grazed yearlings were similar to those in a feedlot at slightly under three pounds a day. These gains would fall whenever the weather became hot as the alfalfa and grass lignified and became less digestible.

The yearlings traditionally gained 300 lbs in 120 days for a season-long average of 2.5 lbs per day.

In the mixed grass/legume pastures, bloat did not become a problem until the weather started to cool in late summer.

---

**Winter Annuals in the Midwest**

In the 1999-2000 growing season at the Linneus Forage Research Station in northern Missouri, where there was adequate fall rainfall, yields for rye in mid-December were 3404 lbs of dry matter versus 1346 for the leading annual ryegrass and 1366 for wheat.

Yields at Columbia in the same year were virtually identical.

Cereal rye maintained its yield advantage over wheat and annual ryegrass through mid-March. However, after mid-March the cereal rye matured much earlier in the spring than did the wheat or annual ryegrass.

Brother Nicholas said after the first couple of years he learned the nuances of grazing alfalfa and never lost another yearling to bloat.

He said *the key thing to watch was the color of the alfalfa when the animals enter a paddock.*

If the alfalfa is a deep dark green you won't have to worry about bloat. It is the light green, fast-growing, immature alfalfa that causes bloat.

He said the color of the alfalfa can be controlled by the rate of irrigation. Alfalfa can be brought to the dark green color by cutting back on the amount of water you put on the paddock.

"If cattle start to blow, hold off on the irrigation until the alfalfa turns dark green."

Irrigation should be done after animals leave a paddock since wet alfalfa is particularly susceptible to causing bloat.

He said that alfalfa bloat was primarily a problem for young cattle. He rarely had any problems with older animals on alfalfa pastures.

While watching so closely for bloat may sound onerous, it is a job that pays better than most. In the 1950s the $300-an-acre net profit alfalfa grazing produced exceeded anything else possible in Utah. He said *direct grazing was at least three times more profitable than making and selling hay.*

At the high altitude Abbey, grazing traditionally ceased on September 1. By that time the yearlings were weighing 1000 lbs and were sold. After the yearlings were removed, the alfalfa pastures were then allowed to regrow until frost.

---

**A combination of fields planted to several species is a good way to sustain quantity and quality while controlling risk. Chains of winter annuals need to be planned early, in the last 10 days of summer or first week of fall in most temperate areas, and paddocks allocated to species intelligently.**

**Dr. Anibal Pordomingo**

After frost, the cows and calves would be brought down from the mountain range to graze out the frosted alfalfa. Once the alfalfa in the pasture was gone, the calves would be weaned onto alfalfa/grass hay.

Brother Nicholas said that care must be taken when calves are weaned and put on alfalfa hay or there can be bloat problems.

He fed weathered or sun-bleached hay rather than green hay in the first few days until the calves became accustomed to the alfalfa.

He said it was critical that the calves never be allowed to run out of hay. Close monitoring was necessary to match the amount of hay on offer to consumption. He said the Abbey used a free-choice hay barn and the hay could be easily dropped into mangers from above. He said such hay barns were extremely labor saving.

He said that by-far *the best stocker cattle performance would come from spring-cut hay. The worst gain was from mid-summer hay cut in hot weather.*

Wastage on summer cut hay was extremely high as the animals refused to eat the heat-lignified stems.

"Those calves would give you a dirty look if you even thought about feeding them summer-cut hay," he said. "They could tell what season hay it was before they took a single bite."

The calves would eat fall cut hay but without the relish they exhibited for spring-cut hay.

He obtained an average daily gain of around 1.5 lbs per day with spring-cut alfalfa/grass hay. Gains would be a pound or less per day if summer or fall cut hay was used.

## Adding Grasses to Alfalfa

Anibal Pordomingo said *the only reason you would want to add grass to an alfalfa pasture would be to reduce bloat and to increase fall and winter dry matter production after the alfalfa goes dormant.*

Anibal found perennial ryegrass, orchardgrass, brome-grass and Reed's canarygrass to be far superior to fescue and wheatgrasses. *The higher the quality of the companion grass the less the chance for legume bloat and the higher winter gains will be.*

For long stand survival, a grazing method that leaves 35 to 40% of the alfalfa as a residual is better than one where the alfalfa is grazed short similar to a hay harvest.

This lax grazing seems to favor bud activation for re-growth and allows a greater portion of carbohydrates to be synthesized in the remaining forage. This acts to spare root energy reserves. Stand life is highly correlated to plant energy root reserves.

Rotational grazing is absolutely necessary to maintain a high level of alfalfa in a pasture sward. The use of 2- to 7-day grazing periods followed by 30- to 40-day rest periods are the preferred strategies during the growing season for maximum alfalfa survival.

There are three types of white clovers: small, intermedi-ate and large. The small type is mostly used in lawns and has little application with beef cattle. Large or ladino white clovers are larger leafed, later blooming and more upright growing than either small or intermediate white clover types.

Under optimal soil fertility and grazing management, ladino white clovers are more productive than other white clover types. However, ladino clovers, dependable reseeders, have fewer stolons and leaves close to ground level. Because of these reasons, ladino clovers have lower grazing persistence.

---

**Anibal's Perennial Grass Choices**
**Ranked by Preference:**
1. **Perennial ryegrass**
2. **Orchardgrass**
3. **Bromegrass**
4. **Fungus-free fescue**

Intermediate white clover is much more persistent under grazing. It typically reseeds better than ladinos, possesses many stolons and leaves at the ground level and produces more forage than the small types.

White clover is very competitive with companion cool-season grasses when seeded together. So competitive in fact that the white clover may smother out the grasses. Consequently, *some graziers prefer to delay clover seeding until the grasses are established.*

The recommended seeding rate for white clover is two to three pounds per acre. Keep in mind that a pound of white clover seed contains 750,000 seeds!

Seeding depth should not exceed 1/4 - 1/2". It is very difficult to adjust a standard grain drill to accurately plant seeds at this shallow depth. Because of the necessity for such shallow planting the seed is often broadcast rather than drilled. When broadcast, seeding rates should be increased by 25%.

Frost seeding in climates where this is applicable is a very effective method in February but less so in the fall. Applications of nitrogen or high nitrogen poultry manure should be avoided in the establishment year as these will favor grass growth and increase competition. This is very bad because you want a legume-dominant pasture for finishing.

The legume percentage should make up 60% of the available dry matter. Such a high percentage is nearly impossible to maintain without a periodic rotation and replanting as the soils become nitrogen saturated and grass dominant.

Established stands of alfalfa or white clover will typically

---

**Fall Grazing Sequence:**
1. **Oats**
2. **Cereal rye**
3. **Triticale**
4. **Wheat**
5. **Annual ryegrass**

produce in excess of 150 lbs of nitrogen per acre per year. Under grazing, very little of this nitrogen is lost and it will build up in the soil.

The decomposition of the underground legume nodules provides a very slow release, stable, nitrogen source. What we do with a forage chain is to utilize this soil nitrogen bank to grow annual forage crops to bump average daily gains up to the finishing level.

**Starting the Forage Chain**

I will start the forage chain with the transition from perennial pasture to winter annuals.

Keep in mind that when we are using natural soil nitrogen it is important that we give a "soil digestion period" after each plow down to allow the soil microbes to release the nitrogen tied up in the plants. *With a perennial pasture the minimum digestion period will be 60 days and with an annual it will be 45 days.* This fallow period will also allow soil moisture to build. This is critical as the late summer is normally the driest period in much of the United States.

A key part of a forage chain is to *use annual forage species with varying maturity lengths to allow for stagger-planting.* For example, cereal rye will mature much earlier than annual ryegrass. This is why you never want to mix species in the same paddock. *Each species should have its own separate paddocks to allow for some (7 days apart) staggered plant-ing in the fall and staggered plow down in the spring.*

---

**Research in Argentina has shown that 50 to 70% of the total forage production of a winter annual takes place in the first growth (first grazing). The earlier the planting, the greater the forage yield at first grazing, and also the greater the chance for developing a good second grazing.**

**Dr. Anibal Pordomingo**

Again, look at the forage chain chart. Cereal rye is losing its finishing quality in March, wheat in April, oats in May and annual ryegrass in June. This should be your sequence for turning them under for replanting to a summer annual.

*Keep in mind that in the spring and the early summer your perennial pasture is of finishing quality, so lingering on a maturing annual is not necessary.* Some Argentine graziers prefer to cut their late spring annual ryegrass for silage rather than graze it out. This silage is then used to supplement stocker cattle on perennial pastures during the summer.

The farther north you are in latitude the more critical it is to plant early to build forage mass before cold weather. These annual forages should be planted in a sequence as well.

In the mid-latitude region (Oklahoma and Kansas), oats and annual ryegrass should be planted in the late summer (August 15) and cereal rye and wheat in the early fall (first two weeks of September).

Oats and cereal rye will be ready to graze 35 to 45 days following planting, wheat slightly later and annual ryegrass at 50 to 60 days later.

*The grazing sequence should be oats first, followed by cereal rye, wheat and ryegrass.* The annual ryegrass can be deferred until the following spring with no loss of quality but the fall and winter grazing of the small grains is necessary due to the possible risk that frost will kill the leaf tops.

Rotational grazing should always be used with winter annuals.

---

**The timing for entering the field for the first grazing of winter annuals is crucial. Too early (less than 60 days after emergence) results in washy, very low dry matter and nutritionally unbalanced feed. Too late results in excessive supply, early maturing of grass and exposure to winter kill.**

**Dr. Anibal Pordomingo**

# WINTER ANNUAL SPECIES

Here's a quick review of the more popular winter annual species and some ideas on how to best use them.

### Annual Ryegrass

This is the fall-planted standard for high quality beef and dairy pasture in the Deep South. In recent years, Midwestern producers in the Fescue Belt are also discovering annual ryegrass for dairying and beef finishing. Cold tolerance and rust resistance varies by variety. Cold tolerant varieties have performed well in northern Missouri.

Annual ryegrass on prepared seedbeds can produce late-fall pasture in the lower South but is generally an early to mid-spring producer elsewhere. This late winter, early spring growth is also true in the Deep South when overseeded on sod-forming warm-season grasses like bermuda.

Spring-planted annual ryegrass matures several weeks later than fall-planted annual ryegrass. Maturity of both fall- and spring-planted annual ryegrass is highly correlated with moisture and temperature stress. Under stress, annual ryegrass will make a seedhead regardless of its height. As a result, when stressed, maturity cannot be delayed with grazing management.

The quick establishment and strong growth of annual ryegrass makes it an excellent rescue forage for winter-pugged areas and frozen-out oat pasture.

Many varieties are strong natural reseeders, and this can be a problem if small grains for grain harvest are to be planted on the same land.

---

**Annual ryegrass should be your winter annual of choice. Although other winter annuals can match its potential in certain times of the year, annual ryegrass has the best and most even quality balance over the entire production cycle.**

**Dr. Anibal Pordomingo**

---

Beef taste tests in Alabama, California, Argentina and New Zealand all agree that annual ryegrass produces an excellent flavored beef, lamb and milk.

Annual ryegrass has the highest carbohydrate-to-protein ratio of the cool-season annuals. This makes it an excellent choice for finishing animals.

Because it will not make a seedhead until vernalized, its grazing can be deferred through the entire fall and winter with little to no loss of quality.

## Oats

In animal forage selection trials, oat pasture is the small grain forage most preferred by cattle, sheep, pigs and chickens.

Its large seed makes it easy to establish with minimal soil moisture. This makes it a good choice for early fall pasture.

In the North, it can be planted in both the fall and spring.

However, in areas north of the Dallas, Texas-Jackson, Mississippi-Macon, Georgia, latitude, fall-planted oat pasture will often freeze out. There are cold-tolerant and high-leaf forage oat varieties but these are not widely available in North America.

Oat pasture is widely used for winter-grazing in east central Texas. Oats in combination with turnips are increasingly being used in the Midwest for high quality, late-spring and early summer pasture.

There are no reports of off-flavors in either meat or milk from oat pasture.

## Wheat

This is the cool-season standard in the Southern Plains. Wheat is more cold tolerant than oats and annual ryegrass but not as cold tolerant as rye.

Grazing wheat actually makes it more cold tolerant. If the grazing animals are pulled early enough, the grazing does not result in lowered grain yields.

Consumer taste tests in New Zealand have identified wheat pasture as producing a noticeable off-flavor in both meat and milk. As a result, wheat pasture should probably be avoided for at least three weeks prior to harvest with beeves.

**Cereal Rye**

Often confused with annual ryegrass, cereal rye is an extremely cold tolerant small grain. Cereal rye will allow a green, living winter feed in most areas of the United States.

The earliest maturing small grain, cereal rye is often used as a winter grazing crop in conjunction with early spring-planted crops such as peanuts, cotton and corn in the Deep South. In beef finishing situations, this early maturity makes cereal rye a much better cool-season companion to warm-season crabgrass than annual ryegrass.

Unfortunately, cereal rye is less palatable than annual ryegrass and the other small grains. When used in a mixed planting, the livestock will consciously select against the cereal rye. Therefore, it is best used in a solid stand rather than as a mix with annual ryegrass.

Because cereal rye will go to seed without vernalizing, it cannot be fall stockpiled for later grazing as annual ryegrass can. Grazing should start when the plant reaches 6" in height.

**Triticale**

Triticale is a genetic cross of wheat and rye. This makes it more cold tolerant than wheat and it is frequently used for winter-

---

**The last grazing of small-grain winter annuals can have seedheads that could possibly compromise a 100% forage program. Therefore, if a field is going into seed before grazing, you will need to mow it high, which will cut the reproductive stems and seedheads formed and favor leaf growth.**

**Dr. Anibal Pordomingo**

grazing in the Pacific Northwest.

As with wheat, there are both spring and fall planted varieties. For grazing you want the fall planted one.

Triticale produces a mild off-flavor in meat but most consumers do not find it objectionable.

This off-flavoring can be eliminated by grazing on annual ryegrass for three weeks prior to slaughter.

Note in the Forage Chain chart (pages 158-159) the increased average daily gain we get in the spring from adding a legume such as vetch or trefoil. These legumes are cheap and easy to grow and should always be included when planting winter annuals.

As you can see from the Forage Chain chart, if we plant early enough winter annuals can make winter a period of easy finishing. What is much, much harder in the summer.

While the drop in animal performance has traditionally been blamed on the heat's effect on the animal, a far greater reason is the heat's effect on the plant.

All perennial forages, with the exception of white clover, lignify and become less digestible when daytime temperatures reach 86°F (30°C). Since this encompasses the majority of the USA, grass finishing in these areas requires that you use warm-season annuals that are more digestible.

## SUMMER ANNUAL FORAGES

Too few graziers in North America have properly appreciated that summer is just as — or more — challenging than winter in producing finishing (and dairy) quality forages. This is because perennial forages (including alfalfa and red clover) lignify at temperatures above 86°F and become less digestible.

Research at the University of Georgia on grazed alfalfa found that the same varieties of alfalfa that would produce over two pounds a day on stocker cattle in June, only produced a pound a day in August due to heat lignification.

This also is why smart alfalfa hay buyers seeking the highest digestiblity hay always specify "spring cut."

Here are some forages to consider for summer use:

## Soybeans

Combining corn and soybeans in a summer forage chain is a good alternative to perennial pasture to sustain summer weight gains.

Coming off of legume/grass perennial pastures in late spring, near-finish-weight cattle would enter the soybean pasture for about a month's grazing. These soybeans will have been planted following the plow down of the early maturing cereal rye crop. Winter annuals and soybeans are very complementary and do not overlap that much in the growing season. Consequently, they make an excellent two-crop rotation.

*Soybeans make a good early planting because as legumes they provide their own nitrogen and are not dependent upon the microbial breakdown to release the nitrogen from the previous crop.*

Soybeans are very efficient water users and require little water prior to pod formation. This is important as late spring prior to the summer storm season is typically a dry period. When laxly grazed, soybeans will produce a finishing level of gain from May until early August.

As with corn, you want to use varieties that produce a lot of leaf and are extremely late maturing as you do not want any pod set. A tropical variety would be best if it can be obtained in your area. These are often called "Florida soybeans" in the USA.

Whereas corn is typically grazed in mid and late summer, soybeans are best grazed in early summer. Crude protein contents of soybean leaves have ranged from 16 to 18% and "in vitro" digestibility from 55 to 68%. In other words, soybeans have a quality similar to alfalfa in late spring. However, unlike alfalfa they do not become less digestible in hot weather.

Soybeans for grazing are typically planted in the mid-latitudes in the first week of May and are ready for first grazing 45 to 50 days later. Planting in rows one foot apart and at an in-row density that is 20 to 25% higher is recommended.

Depending upon moisture availability, two regrowths and grazings can be expected every 35 to 40 days. A fourth grazing could be possible but will conflict with the planting of winter annuals. Where a winter annual is not desired, some graziers in Argentina have been able to get five grazings out of a single planting.

Being a bushy type plant, *it is important to only take about half the leaves on offer for a fast regrowth* although in research trials up to 70% of the leaves have been removed without killing the plant.

Some graziers will plant corn and soybeans in four rows of one and four rows of the other. The corn provides the necessary carbohydrates and dry matter and the soybean complements with protein and digestible fiber.

They will then graze the two plants together in a "slice" made with temporary one-wire electric fence. However, having some separate paddocks of corn and soybeans is an excellent

---

**In Staunton, Virginia, James Wenger prefers a pasture mixture of alfalfa and orchardgrass. He is slowly renovating all of his paddocks from fescue to this mix.**

**His renovation schedule is to burn down the fescue in the fall with herbicide and plant it to cereal rye. This provides both late fall and early spring grazing.**

**In mid-May, these cereal rye acres are planted to brown mid-rib sorghum-sudan for summer grazing.**

**In August, the sorghum-sudan is sprayed out and the land is planted to the alfalfa/orchardgrass mixture. He uses 10 lbs of alfalfa and 8 lbs of orchardgrass seed per acre.**

idea as it will help prevent grazing the corn too early before it has reached its maximum grazable biomass.

For planning purposes, we should figure that it will require twice as many acres of soybeans to finish a set of steers as it will with corn due to their lower biomass. Soybeans also have a lower average daily gain than corn but one high enough for finishing.

Of course the premier summer annual for finishing is green leaf corn. Once the stagger-planted corn is ready to graze we can switch our finishing cattle from the soybeans to the corn on a full-time basis or on half-a-day in each.

The partial day graze can help us to better utilize the soybeans and help stretch the corn crop. However, if the cattle are to be harvested in 20 days or less, it is best to leave them on the corn full-time as the gains will be higher.

## Green Leaf Corn

In a green, leafy stage there is no better forage for finishing beeves during the hot summer months than direct-grazed corn. The longer and hotter the summer in your region, the better direct-grazed corn will look to you.

Digestibility of the whole plant before tasseling are usually above 70%. This compares favorably with the digestibility of corn-based feedlot diets. Crude protein ranges between 12 and 16% depending upon soil fertility and maturity stage and dry matter content is between 20 and 35%.

Of course, the real kicker is a soluble carbohydrate content above 25% and a highly digestible fiber. The net result is an almost perfect plant for finishing cattle.

While with all other forages average daily gain falls as the plant matures, with corn the gain is high and steady right up until the grain develops. This high gain period is normally at least two months in length.

By high gains, I mean between two and three pounds per day in hot, sultry, July and August! And, this gain is almost all fat! Incredible!

Of course, the two drawbacks to corn are that it is an annual, and it has no regrowth following grazing. To me this is offset by the fact that *the amount of dry matter produced in 45 to 50 days of growth is equal to or higher than what a perennial warm-season grass could produce in a whole season.*

And, unlike almost all other summer grown forages, it will hold its quality over a relatively long period of time. This makes it an almost fool-proof forage for successful summer finishing.

Green leaf corn grazing should perhaps be considered as a summer stockpile grazing program. First let the plant grow and accumulate dry matter and then ration this stockpile out as you need it with temporary electric fences just as you do with a fall grown, winter-grazed cool-season perennial.

With grazed corn, always use the longest maturing silage varieties, as you are not interested in grain production. These slow maturing varieties will be those designed for Florida and the Gulf Coast region. Some graziers prefer the older, open-pollinated varieties due to their superior drought tolerance, long green season, higher protein and digestibility.

Corn should be planted with 20 to 40% more seeds within the row than that recommended for grain. However, row width should remain the same so tillage can be used for weed control.

---

**Steers initially weighing 700 to 800 lbs averaged nearly four pounds per day for 90 days on grazed corn in a Mississippi trial. The cost of gain was estimated to be one-fourth of a traditional feedlot ration. This is an excellent way to harvest marginal corn crops.**

**A 2003 trial at Mississippi State, grazing the corn netted $157 an acre — far higher than the profit potential of a machine harvest for grain. Total machinery used for the corn grazing trial consisted of a tractor, no-till drill and spray rig.**

Often tillage is not necessary in the first year after coming out of perennial grass but normally it will be necessary in subsequent plantings. I will discuss organic no-till possibilities in a moment.

Some graziers do not till the row centers and some actually plant crabgrass in the middles to thicken the stand. This is of dubious value as it will greatly lower the total dry matter yield that a weed-controlled corn could have produced. ***An alternative to tillage is to use small, fresh-weaned lambs for in-row weed control.*** The key word here is small. These small lambs will typically only eat the bottom row of corn leaves and will concentrate on the crabgrass and other highly digestible weeds. Since the tilled ground will be free of sheep parasites, this is not a bad place to put them.

The dry matter range for corn can range from 3000 to 12,000 lbs depending upon soil fertility, moisture and cultural practices such as in-row weed and grass control.

We want to ration this standing stockpile with strip grazing just like we do a standing grass crop in winter. Otherwise the animals will knock down and sleep on more feed than they consume.

Anibal calls these temporary allocations "slices." I think this perfectly describes what we are doing.

Think of the corn crop as a cake. We are using a one-wire

---

**On the 1000-head finishing ranch he operates, Anibal prefers to rent tillage equipment on a day rate from a neighbor rather than own it. The ranch's current equipment inventory consists of one small Japanese pickup, a small flatbed trailer, a small bucket shovel attachment and a one-bale hydraulic fork.**

**He is considering buying a mower (maybe) and a 120 HP used John Deere tractor. The ranch has only one full-time employee and all occasional labor used is contract labor.**

electric tape (tape is more visible) to slice off daily allocations of cake for our cattle. We want these slices to be of a size that will last for one day, or even better, a half-day.

*Half-day slices will produce a higher average daily gain and will finish your cattle faster.* This higher gain is the result of the animals being able to select a diet higher in leaves.

The risk in a half-day slice is that you can accidentally under feed the cattle, and they are forced to graze stems as well as the leaves. Stem grazing will greatly lower the average daily gain.

For maximum gains, at least half the stem of the original plant should still be standing when the cattle are given a fresh slice of corn. If only 1/3 of the original stem remains, our animal gains will be significantly less.

This remainder does not have to be wasted as it will be good cow feed. This means you will have to back fence the corn to keep the two groups separated.

By placing a movable water tank under the back wire both herds can be watered from the same tank. Just make sure you have enough water pressure to keep the tank full.

While the back fence does not have to be moved as frequently as the forward fence, it needs to be moved frequently enough to keep the water close enough to the finishing cattle so that they drink individually rather than in groups.

Keep in mind that the more efficient we try to become in our grazing, the greater the risk of accidentally under-feeding the finishing cattle.

At the price of today's grass-finished cattle, erring on the side of plenty is the wise economic decision.

Most graziers use their ATV, pickup or tractor to knock down a swath of corn for the placement of the crossfence. This is important because we want the cattle to easily see the fence and not accidentally stumble into it as this will make the cattle shy and tentative in their subsequent grazing. This would be a reflection of psychological stress we have unwittingly introduced to the cattle.

Remember with finishing cattle we want no restrictions on the animals' grazing from either a quantity, quality or stress standpoint.

### So, how do we figure the stocking rate?

A finishing weight animal will eat about three percent of its bodyweight a day. For example, a 750 lb yearling will typically eat about 22 to 25 lbs of dry matter a day.

If we have 8,000 lbs of corn dry matter in our paddock and are only going to graze half of it, or 4000 lbs, we can divide 4000 by 25 (the dry matter consumption) and this will give us a stocking rate of 160 beeves needed to graze an acre in one day.

While there are all sorts of mathematical formulas you can use to estimate standing corn dry matter and interpolate correct stocking rate, I doubt that many of you will ever do this.

Most of us will initially allocate the corn by trial and close observation. Again, wasting corn is far less expensive than under-feeding finishing animals.

We typically want to first enter the corn field when the corn is about four feet high. Corn can be grazed shorter than this with no problems but overall dry matter yield will be reduced.

At the four-foot-height initial graze level, this will give us 50 to 60 days of green leaf grazing before grain appears. At two pounds a day this is between 100 and 120 lbs per head.

Now, here's a problem. Because there is no subsequent forage that equals corn's quality in late summer and early fall, we want to harvest our finishing animals straight off the corn.

*While there is no rumen adaptation lag going to corn, there is leaving corn and going to pasture.* As a result, gains will collapse and possibly stop altogether for 15 to 20 days or more. This will probably destroy the eating quality of the meat. Therefore, *it is critical that the animals put on corn be of a weight capable of finishing before the corn runs out.*

For example, if we have determined from a steer's dam

weight records that he will probably finish at 1100 lbs. He needs to be weighing 1000 lbs when he goes on the corn if he is to finish in 50 days. Or, we need to have a longer corn graze.

We can do this by stagger planting the corn.

Rather than planting all of our corn at the same time, we want to plant the field in smaller sections at 15- to 20-day intervals. This corn planting should start in early spring and possibly continue until the first week of summer.

*Typically, grazing can start 45 to 50 days after planting.* This could give us as long as a four-month corn grazing period and even longer in the Deep South.

Corn plants will stay green and highly palatable until frost. This allows a late-planted corn to finish beeves in October and November when no other forage is capable of producing finishing level gains.

This necessity for staggered corn planting is why we never mix small grain varieties and why we want to use species with different maturity dates. For example, cereal rye will mature faster than triticale, triticale faster than oats, oats faster than annual ryegrass. This allows us to plow them down in a staggered manner for subsequent staggered corn planting.

---

**To Prevent Fall Scouring:**

■　　Feed good quality hay or mixed pasture silage free choice with no restrictions.

■　　Mow and graze the windrows a day later.

■　　Plant species with a slower initial growth (oats or wheat vs cereal rye).

■　　Plant species that stockpile with good quality (annual ryegrass vs cereal rye).

■　　Reduce or eliminate nitrogen fertilizer use at seeding time.

■　　Use no-till planting if possible. This results in less soil nitrogen being available at the early stage.

**Dr. Anibal Pordomingo**

---

Of course, keep in mind the necessity of a 30-day "soil digestion" period before planting the corn. This digestion period is necessary to allow the soil nitrogen tied up in the plowed-down crop to be released for subsequent reuse by the following crop. Crops plowed down while still green have a much faster nitrogen release.

In drier climates, a winter fallow may be necessary to accumulate enough moisture for the corn to grow well all summer. Irrigation greatly expands the reliability of an all-forage finishing program.

Staggered grazing of the corn also allows for a staggered planting of the subsequent small grains for winter grazing. By disking the corn ground rough immediately after the cows clean up the paddock, we can then leave it fallow for summer moisture accumulation. This will greatly help the late summer- and early fall-planted winter annuals get established before cold weather.

Such double cropping options as I have discussed will not be as successful where soil nitrogen (organic matter) levels are low. Therefore, rectifying this problem with plowed under legume crops or going to a multi-year legume/perennial pasture phase first should by your priority.

### Equipment Needs

Anibal said that a winter or summer annual forage chain actually requires very little equipment.

A 120 HP tractor should suffice for a 600- to 800-acre grass-finishing farm. This same tractor can be used for planting and cultivating. You will need a seeder capable of planting small grains, perennial grasses, summer annuals and legumes.

If using no-till, you will need a no-til seeder capable of applying starter fertilizer at planting.

For tillage tools, you will need a chisel plow and a heavy disk drag harrow. Thanks to the natural weed control you get from a grazing operation tillage requirements should be minimal.

Here are the two most common methods for preparing and planting small grains:

## A. Conventional Minimum Till

One pass of a chisel plow or heavy disk harrow. Anibal far prefers a chisel plow. If a green manure is to be plowed in for nitrogen this should be done after chisel plowing in the summer prior to planting.

One pass of a disk harrow.

One pass of a small-grain seeder. If your soils are low in organic matter, nitrogen and phosphorous should be applied.

Anibal said it is far better to use natural nitrogen from plowing down a leguminous pasture or planted green manure with finishing cattle to avoid artificially raising the nitrogen level.

## B. No-till

Graze down and/or kill the previous crop with Roundup. With high stock density grazing (brood cows) it is often possible to avoid using Roundup at this time.

No-till plant in late summer with a no-till planter with starter fertilizer of nitrogen or nitrogen/phosphorous (depending upon soil test).

Apply Roundup the day after planting is completed. (This can also be done two days prior to planting.) This Roundup application can also often be skipped by using cows to graze off the weeds.

## What about Organic No-till?

An organic no-till program is indeed possible in a perennial pasture/annual crop rotation. However, Anibal said he thought this should only be attempted by master grass farmers who are competent in both Management-intensive Grazing and organic soil building techniques.

Anibal said what makes organic no-till possible in a forage chain situation is that the managed grazing during the multi-year perennial pasture phase greatly eliminates most annual weeds that are a problem with cropping.

One of the reasons weeds are such a problem with conventional no-till planting is the use of heavy amounts of artificial fertilizer. However, if legumes are a part of the forage base, as they should be, the need for extra artificial nitrogen will be eliminated. Similarly, high organic matter soils tend to be naturally higher in phosphorous.

Under grazing, even at the highest stocking rates, the removal of nitrogen and phosphorous does not exceed 10% of the total nutrient collected in the above ground portion of the plant. From the fraction of the plant that enters the animal's digestive tract, less than 8% of the nitrogen and 15% of the phosphorous are retained as animal meat and bone. Most of the nutrients of the pasture eaten are returned to the pasture through the feces and urine.

Anibal said Argentine research has shown that, on average, nitrogen extraction in grassfed beef does not exceed four percent of the total nitrogen offered to the animal. In contrast, grain crops can take from 60 to 95% of the available nitrogen.

**The big problem with no-till in a grazing situation is dealing with soil compaction.**

Traditionally, this is solved by periodic subsoiling with a chisel plow. This is typically done every three to four years on annual pasture and every five to six on perennial pastures.

You do not want to subsoil all of your annual ground at once as subsoiling makes the paddock particularly prone to pugging in wet weather. However, soil compaction tends to be a soil specific problem. In general, the higher your soil organic matter is, the less of a problem you will have with it.

Plowing down leguminous forages like vetch and soybeans and then planting the ground to a legume/cool-season perennial grass for several years are excellent ways to build both soil nitrogen and compaction-resistant organic matter.

You cannot go from conventional cropping directly to

organic no-till. You must go through the perennial pasture phase first as this is when you build the soil nitrogen reserves that allow you to stop nitrogen fertilization. This phase also allows you to get your weed problems under control with controlled grazing.

An organice no-till program would not be successful in an all-annual forage chain.

## Considerations Before Going No-till

Here are a few things Anibal said to consider before introducing no-till planting in an organic pasture system.

■ There should not be a pre-existing heavy weed load.

■ You should have the Management-intensive Grazing infrastructure in place and know how to use it.

■ Forages and their proper sequence should be well understood and the weak points in your growing season and forage quality identified. (Quality, growth rates, survival under compaction and competition, nutrient needs.)

■ A plan for organic matter and soil cover/soil mat increase should be written out. (Organic and green manures, stubble build up.)

■ Necessary soil nutrient sources should be identified. (Nitrogen fixing legumes, amount of soil sodium, phosphorous, potassium and calcium.)

■ Soil limitations must be clear and removed if possible. (Soil pH, mineral ratios, depth, hard pans, organic matter, water holding capacity.)

■ Soil compaction should be assessed for every field and broken up if it exists. You must know how easily your soils are compacted.

Farms that have a history of conventional farming cannot be transformed to organic farming in a day. It takes time to eliminate the residual effect of pesticides and chemical fertilizers. In most cases, the decision to go organic will require the restructuring of the entire production system.

## Crabgrass

One of the most over-looked warm-season annuals is crabgrass. No doubt, crabgrass' primary problem is its long association as a major lawn and rowcrop weed. While there is nothing particularly sexy about crabgrass, as a warm-season forage in sandy natured soils it is superb. In Nebraska, I know of at least one grazier who plants crabgrass in the row middles of his grazed green leaf corn.

With proper grazing management that emphasizes leaf-only grazing (leader-follower), average daily beef gains from crabgrass can approach those of cool-season annuals such as wheat, oats and annual ryegrass. *On lush stands, gains of 1.98 to 2.87 lbs per day have been recorded. On average, crabgrass will produce an ADG that is 50% higher than on bermudagrass.* These gains decline below that needed for finishing after late August in most locations.

Not only is the crabgrass a strong natural re-seeder but it double-crops well with all the popular winter annuals — cereal rye, oats, wheat, annual ryegrass, triticale.

What is too little understood about the grass is that it thrives on tillage. While crabgrass will often volunteer in behind most tilled winter annuals on its own, it does so much faster and thicker when the soil is tilled three inches deep and rolled smooth at the end of the winter grazing season.

The 21-year average for crabgrass after winter pasture in Oklahoma research at the Noble Foundation was 3,075 lbs of dry weight forage. The-beef-to-forage conversion is about 10-to-1 for yearlings and 6-to-1 with lightweight stockers on high quality stands. This means around 300 lbs of beef gain per acre from a 90-day summer grazing season.

While crabgrass is susceptible to dying in summer droughts, the standing dry forage is still of relatively high quality and suitable for stocker cattle.

Using a leader-follower grazing system whereby steers were followed by bred heifers, the average daily gain for the steers in June was 2.90, 2.90 in July and 2.42 for the first 15 days of August. The total rainfall for the last two months was only 2.5 inches.

The second grazer heifers who were forced to clean up the grass residue following the steers gained 0.77 per day for the entire trial. The heifers were grass fat going onto the pasture and weighed an average of 968 lbs.

Figuring tillage and fertilizer cost the cost of gain for the steers was 12 cents per pound of gain and seven cents for the heifers. Not bad for a weed!

### Eastern Gamagrass

While not an annual, this is the only warm-season perennial I have found that is capable of producing finishing level gains for longer than 45 to 60 days. Argentine research has found that gamagrass can produce two pounds a day gains from April until late September (North American equivalent months). This is equal to the quality period length of perennial ryegrass and white clover.

Gamagrass is a native warm-season bunchgrass that is a close genetic cousin to corn. It has wide leaves that closely resemble those of corn's. *The plant grows to a height of six to eight feet and will produce two to five tons of dry matter with no supplemental nitrogen.* The plant's adaptability range extends from Mexico to Nebraska.

Its preferred soil group is a wet-natured highly fertile soil. As such it might be the plant of choice for winter-wet, pug-prone, river bottomland. A take-half, leave-half, high residual rotational grazing program is absolutely necessary for this plant to survive. The plant spreads primarily through underground rhizomes and its seed production is very low.

The primary drawback to gamagrass is that it is difficult to get established. Grazing must be deferred altogether in its establishment year. It is also very susceptible to weed competition in its establishment year. This slow initial establishment and the necessity for corn quality soils has slowed the spread of this very high quality forage in North America.

**Feeding Hay in Summer**

Stocker cattle's gains on perennial pastures can fall to less than one-half pound a day in the August dog days of summer.

These low gains are due to the lignification of the forage due to the heat.

All perennial forages except white clover lose their digestibility when the daytime temperature exceeds 86°F.

These low gains can be increased by the feeding of leguminous hays or silages cut during the cooler period of late spring and early summer.

In Argentina, Anibal said grassfed beef finishers will keep free choice alfalfa hay available to their stocker cattle for as much as ten months of the year to buffer changes in forage quality and dry matter.

*Well produced hays from pure legume stands complement grass monocultures (annual or perennial ryegrass, annual winter foragers).*

Anibal said the two best hays to feed to increase summer gains are pure alfalfa hay or hay made from winter annuals such as oats or annual ryegrass in combination with vetch.

Vetch should always be included with winter annuals to increase late spring forage quality for either grazing or hay. Winter annual hay without vetch will typically only produce half the average daily gain of that with vetch.

Both of these hays when cut in late spring are capable of

> **Quality hays are produced from the spring alfalfa-rich pastures that are cut before 10% of the alfalfa blooms. Frequently, this cutting stage produces less quantity than at mid-blooming, but the quality will match well with the winter annual forages. Such high quality hays are a way to maintain animal performance in late fall , and early and mid-winter.**
>
> **Dr. Anibal Pordomingo**

producing average daily gains of 1.5 lbs per day or considerably more than late summer perennial pasture.

Non-legume grass hays do not have enough forage quality to increase summer pasture gains and should be restricted to dry cow use.

## Using Silage in Summer

Leguminous silages are superior to hay as a summer gain enhancer and in fact should be fed in summer rather than winter if the stocker cattle are being grazed on winter annuals or stockpiled cool-season perennials.

This is because summer pastures are typically high in dry matter due to heat dessication and so the higher moisture of silage is not a drawback.

However, this is not the case in the fall, winter and spring when forages are lush and low in dry matter. In these non-summer periods, the high dry matter of the hay will be a major benefit to the animal's gain performance.

Green leaf corn silage (cut prior to ear development) makes an excellent summer and fall supplemental forage but is still inferior to legume silage with protein-loving stocker cattle.

All silages preserved by removing the air with a vacuum pump are superior to tractor packed silages because less of the carbohydrate portion of the forage is consumed by the heat endemic in tractor-packed silages.

## Irrigation

Irrigation is most cost-effective when it is used to supplement natural rainfall. It is too costly to use all the time.

A little water often is far better than a lot infrequently.

---

**Offering high-protein low-carbohydrate silages on lush high-protein fall or spring forages should be avoided. It will worsen the protein/energy imbalance of the diet and will not improve energy intake.**

---

Best to put on 1.5 inches every six to seven days. It takes the first inch to get the ground wet. The next half inch is what makes the grass grow.

Using small amounts of water per irrigation prevents pasture pugging. Livestock should be moved off the irrigated area during times of excessive natural rainfall to prevent pugging.

The biggest mistake with irrigation is trying to cover too much land with too little water. Better to do 50 acres properly than 100 acres halfway.

***The goal is to keep the ground moist but not wet.***
Overwatering makes grass roots short. Keep soils in optimum moisture range.

Fertilizer rates have to be increased over those on non-irrigated land to maximize profitability.

Plant the most productive perennial, cool-season grass you can find. However, a minimum of 10% of the farm should be in annuals to keep animal performance high in hot weather.

Perennial ryegrass will survive very hot temperatures with irrigation. However, long rotations (45 days or more) or total grazing deferral will be necessary when daytime temperatures are in excess of 90°F (32°C).

Try not to conserve feed from irrigated pasture. Always try to run it through a high-value-producing animal. It's better to bring in stock from non-irrigated areas than to harvest as hay or silage.

Irrigated pasture is most profitable when used for dairy and/or grassfed beef and lamb finishing.

Shelter belts are very important to save water. Plant trees 18 to 20 feet deep on the perimeter of irrigated area to reduce wind evaporation.

# CHAPTER 10

# Laying the Foundation in the Soil

Perhaps it is time now for a gut check. How do you feel? Overwhelmed? I know I was the first time I was presented with a year around forage chain.

Here's some advice:

Whenever you have a big job to do, break it down into a series of jobs and then do these jobs in such an order that success in one leads to success in the next.

If you get these jobs out of sequence, you lose the stabilizing, self-reinforcement of building the base of the pyramid first.

Irish grazier, Michael Murphy, is probably one of the most financially successful grass farmers in the world with an annual net income of around a million dollars. He has done this by leasing financially failing farms and systematically turning them around with managers he has personally trained and cattle he has selected and bred to do well on pasture.

The key word here is systematic. He said you can develop a systems approach to financial success with a very small farm that can be scaled up and repeated again and again.

Murphy is a big believer in the necessity of a first-things-first approach. Here is the sequence he told me that has worked so well for him.

Murphy's very brief advice is in the bold face type. My added comments are in the regular typeface.

1. **Do whatever it takes to grow a quality pasture.**

With grass-finishing this means a highly leguminous pasture. Do whatever it takes in mineral fertilization to grow either alfalfa or white clover depending upon your climate. Realize that the high legume percentage required for finishing will probably require periodic renovation and replanting. Use the planting of seasonal annuals to plow in lime deeply to offset subsoil acidity. Use annuals to solve your biggest problems first.

2. **Use only animals genetically selected for pasture production.**

Finishing is a thousand times easier with mid-sized, easy fattening breeds. Only buy in seedstock from graziers specifically selecting for grass finishing. Be very careful selecting genetics outside of the English breed gene pool. Weigh your cows and sell those weighing more than 1000 lbs.

3. **Get the animals' production and reproduction in sync with the grass growth curve.**

Time your finishing to when your pastures produce their highest average daily gain. With steers this will normally be the spring of their second year. Time your calving to surplus periods when the cow's increase in pasture consumption at calving will help maintain the pasture quality for your stockers and finishers. As much as possible, use your cows as grass scavengers.

4. **Cut labor costs per unit of production.**

Farm out to others all hay and silage making. Keep in mind that contract grazing allows you to buy both grass *and labor* on a per head basis. These costs stop immediately when you retrieve the animal.

I would recommend that you use custom graziers for cows and bred replacements but not for gaining animals except perhaps on species like winter annuals where the inherent quality of the pasture can offset a lack of grazier skills.

## 5. Move up the value chain as rapidly as possible.

It is more important that you know where you want to go than where you currently are. If you are a cow-calf producer and plan to eventually direct-market your own grassfed beef, start on creating that future with the bulls you buy and the calving season you choose.

Always work from wherever you are toward the consumer. Remember, you will not be completely out of the commodity business until you own your own customers.

## 6. Reinvest profits in high return areas.

Believe it or not this is the toughest one to follow. Most of us want to reinvest profits in whatever made them. In other words, in expanding what we are doing. However, this is never the highest return. The highest return is time spent learning new knowledge that most people don't know. This is the true source of competitive advantage.

Until you have a gourmet product, your highest return will come from learning production skills. Once you have a gourmet product, your highest return will come from learning marketing and finance skills.

Keep in mind it is far easier to find people skilled in marketing and finance than it is people skilled in the knowledge of how to grass-finish cattle. As a consequence, to speed things up you may want to hire in these other two skills.

Is this the sequence you are following?

Too many I fear are thinking of jumping over the first four and going straight to number five. The problem with this is poorly selected animals, grazed on poor quality pastures are going to produce a poor eating quality product. And, if your initial products are of low quality (tough, too lean, poor flavor) you are dead in an industry that primarily relies on word-of-mouth recommendations for new customers.

It takes a very unusual North American to understand that the base of a production-for-profit pyramid is the pasture and not the animal. Profit is not a function of production it is a function of production margin. *Nothing produces as much margin per unit of production as direct-grazed forages.*

Culturally, this is alien to us. In North America, we have never learned to "trust" pasture. Whenever we have been faced with the need to grow an extra-high quality pasture, we have turned to grain.

Part of the problem has been confusion over the fact that good pasture advice for beef cows — the dominant grass-based enterprise — is not good advice for dairy cows or pasture-finished beeves. Beef cows make money by adding value to an undervalued resource. In contrast, dairy cows are a value-producing resource in and of themselves.

A beef cow returns only a tiny smidgen (six percent) of the energy she eats in salable production each year. Over 9% of the total energy required to produce a pound of edible finished beef goes to just keeping the cow alive. As a result, for a beef cow to be profitable the feed she eats has to have a very low monetary value. In fact, ideally, it should have a negative monetary value. (By negative monetary value, I mean that if the cow was not there you would have to spend out-of-pocket money to control the vegetation.)

*Normally, beef cows predominate in climates where the feed resource is plentiful but of low quality.* The beef cow can buffer the lack of forage quality for herself and her calf with the fat on her back. Over seventy percent of the world's

---

**Natural Soil Nitrogen Builders**
1. **Plowed under vetch or soybeans.**
2. **Multiple year period in perennial legume/grass pasture.**

    **Remember, artificial nitrogen can cause daily gains in finishing weight cattle to fall.**

beef cows are located in the tropic and sub-tropic regions where grass is seasonally abundant but of relatively low quality.

In temperate humid climates, beef cows are best used for a pioneer species (bush-basher) in the development phase of a farm or ranch or as a seasonal grass quantity and quality controller on a developed farm.

Since beef cows double their forage consumption at calving, these late spring calving beef cows add the necessary "put" in a "put-and-take" harvest system to keep the pasture quality high for other higher value species.

Thanks to their ability to cushion feed consumption shortfalls by utilizing back fat, beef cows respond well to no-input, rangeland, no-tillage, no-machinery-but-a-wheelbarrow system.

However, there is nothing "natural" about a high Select/low Choice steer in January. You can't remain a rancher and do this. You can if you are willing to become a grass farmer.

I shocked a group of graziers the other day by saying that in 20 years all grass finishing operations would be on Class 1 farmlands. The same quality soil that will grow 185 bushels of corn is the same quality soil that will best grow the necessary "unnatural" finishing quality forages, and a grass-finished beeve will be highest value crop you can grow there.

New Zealand's famous Waikato Dairy District and Argentina's beef "finishing zone" are those countries best land, not its worst. It will be the same in North America once we gain confidence in pasture.

I have seen a lot of grass dairies fail because they just weren't on high enough quality land to start with. We will see the

---

**Plentiful soil calcium is absolutely necessary for tender beef.**

**Alluvial soils tend to be very deficient in sodium. This makes them particularly susceptible to legume bloat.**

same with pioneer finishing operations. Steep hillsides are for sheep — not finishing cattle. Anibal continually talks about the necessity for "easiness" in grazing.

*We cannot randomly choose enterprises with no relation to the landscape and soil quality. High-value-producing enterprises require high quality soils to grow high quality forages.*

The driver of all pastoral systems is soil nitrogen. This reserve of plant-available nitrogen is called the nitrogen "sink" or pool. Plant-available nitrogen comes from the breakdown of soil organic matter by micro-organisms. The greater the nitrogen pool in the soil the higher and more stable the pasture production will be even if commercial nitrogen is used as the primary N driver.

Keep in mind, the commercial nitrogen you buy is not in a plant-usable form. It merely speeds the microbial breakdown of pre-existing soil organic matter to make plant-usable nitrogen. At some point the soil becomes virtually exhausted of organic matter and nitrogen no longer works well enough to be cost-effective. Whether you are certified or not, everyone in agriculture is "organic" farming.

The beauty of a grazing system is that clovers, which increase animal performance and production by increasing forage quality, also increase forage quantity by increasing the soil's reserve

---

In low organic matter soils, growing and plowing under a crop of vetch is recommended before planting perennial pasture.

Adding vetch to winter annuals can increase ADG in early spring by one-half of a pound.

After plowing under green perennial pasture, a digestion period of 45 days should be allowed before replanting.

After plowing under green annuals, a digestion period of 30 days should be allowed before replanting.

of nitrogen. However, they do not do this at once.

What is misunderstood about the clover/nitrogen cycle is that the majority of the soil nitrogen comes not so much from the nitrogen-fixing nodules of the clover as from the return of the plant's high level of nitrogen (protein) in dung and urine.

This is why trying to grow annuals from the nitrogen produced by a companion legume in low-organic-matter (low nitrogen reserve) soils is always doomed to failure. *This year's crop lives on nitrogen reserves built in the years before.*

All cool-season perennial grasses require a companion legume to be of finishing quality. Therefore, most pasture fertilization of lime, phosphate, sulphur and trace minerals is to grow the legumes. (Many soils that have high pH's are low in calcium.)

However perennial ryegrass — the ultimate cool-season perennial — doesn't allow us to cheat on pasture fertilization as this particular grass requires almost the same high levels of calcium, phosphorous and sulfur as a legume. On non-irrigated areas it also requires a high level of organic matter (and long grazing rest periods) to make it through the dry periods.

As a result, perennial ryegrass is not a good species to start with, but an excellent one to plan to use once your soil's calcium, phosphate and organic matter have been built up. Again, it should be part of your plan even if your soils are not ready for it today.

I saw on the news recently a robot surgical knife that would allow a highly skilled surgeon to operate on a patient while separated by as much as 10,000 miles. As amazing as this invention is, when the robot gets through with you it is going to take you just as long to heal and get over the surgery as it ever did. Your soil is the same way.

Modern row cropping is the soil's equivalent of having been on the receiving end of a mugging by Jack the Ripper. Your soil is in a state of shock. Many of these soils have so little soil life left that they will need to be re-inoculated with micro-organisms. Spreading compost on these shocked soils is one good way to bring them back to life.

Large inputs of fertilizer (particularly lime) are required in the early stages of bringing a dead soil back with seemingly little result. This is the phase when we are putting our money in the bank (the soil's nitrogen bank) so that we can draw interest to live on later. Like most investments, this will result in our having a diminished or possibly negative cash flow while we are getting the farm up to speed.

This lag between expectations and results is why grazing is frequently said to bring on bankruptcy. No farm can turn around biologically (or financially) on a dime. You can't wait until you are ten feet off the ground to pull the ripcord of even the best designed parachute.

Soils that are low in nitrogen will be legume and weed dominant until the soil nitrogen level rises enough for the grasses. This will take on average, two to five years in a humid climate.

Keep in mind the first thing nature wants is the soil surface covered.

***Weeds might be thought of as Nature's scab.*** Let them do their job and graze the weeds with cows. Mowing the weeds will both keep their quality high, reduce their seed production and help produce ground covering mulch and litter.

At some point, if you keep fertilizing, pasture production starts to increase and soon hits its stride increasing on almost a one-for-one, input-to-output ratio until optimum soil nutrient status has been reached.

---

**To help reduce legume bloat, always drill grass and legumes in separate rows so the animals can choose and interspecies competition is lower.**

**Always plant different species of winter annuals in different paddocks. Do not mix them together.**

**There is no rumen adjustment period going from cool-season grasses to green leaf corn.**

Once this stage is reached more fertilizer does not grow that much more grass, and fertilization can drop to just an occasional maintenance application.

While it is true that grazing and manure can do some wonderful things in increasing soil organic matter, it will never create or replace deficient soil minerals.

From what I have seen, ***the major difference*** between successful graziers and unsuccessful graziers is whether they have made the investment to correct soil mineral deficiencies.

Yes, in a desert you want to try to feed the minerals directly to the animal, but in an intensively managed pasture situation you want to feed them through the pasture via fertilization. A healthy, legume-rich pasture *naturally* produces healthy animals. And, as Jo Robinson has pointed out, we are what our animals eat.

Legumes are your key indicator plant for mineral balance and deficiency. If you have adequate moisture and you can't grow legumes, there is a soil mineral reason. Find out what it is and fix it.

Getting through points one through four is going to take a minimum of five years for the best graziers. Are you willing to invest *the time* it takes? Most people aren't.

However, profits don't come from following most people. ***Lasting profits always come from doing those things most people are unwilling to do.***

# CHAPTER 11

# Low Stress for Tender Beef

I received a call one late fall day from a West Coast meat vendor complaining about the "gamey" flavor in some of the grassfed beef he was distributing. He said this was turning off a lot of his customers who bought the product.

I explained to him that this off-flavor had nothing to do with grass but with how well the cattle were fed and handled. In other words, it was a management problem and not a cattle or grass problem.

"Why do we seldom have this problem with grainfed meat but it is endemic with the grassfed product?" he asked.

Good question. Here's the answer.

Gamey-ness or off-flavor and an often accompanying bad smell is related to physical exertion and stress on the part of the animal in its final hours.

We call it gamey because it is similar in the taste and smell of venison from deer that was not killed quickly.

Animal behaviorist, Bud Williams, said his father taught him when they were hunting deer for table meat during the Depression that the best eating venison would come from a deer shot while it was asleep in its bed.

Stress creates a dark colored meat with a coarse texture and a reduced shelf life due to an above normal meat pH. The gamey flavor is the result of a combination of adrenalin and bacteria, which grow rapidly on high pH meat. These bacteria produce the off-putting, rotten meat smell. High pH meat also sheds moisture rapidly under cooking, producing a dry, hard to swallow, tough meat that will not lose its pink center despite extensive cooking. This last factor is particularly irksome to consumers who like their meat well done without any pink.

Live muscle has a pH of 7. Meat with good visual appeal and good eating potential will have a pH of 5.3 to 5.7.

Meat from stressed animals will not fall below 5.8 to 6.9. Such high pH meat cannot be vacuum-packaged as it will rot in the package.

Stress suppresses the endorphin and immune systems, triggers alert mechanisms, increases muscle oxygen consumption and muscle glycogen degradation and lactic acid buildup. Muscles darken (lack of oxygenated blood) and become rigid.

Hypoxic muscles take longer than normal to return to a relaxed, oxygenated stage. *If animals were stressed prior to slaughter, muscle relaxation after rigor mortis takes place only partially* and the dark cutter syndrome develops.

Management-intensive Grazing and minimum-stress handling strategies provide stocker/finishing programs a good opportunity for training steers to deal with people and routines. Teaching cattle to deal with the unknown and not be stressed is relevant at loading, trucking and slaughtering time. Low stress at slaughter is highly correlated with beef tenderness.

Built-in stress reduces rate of gain, which in turn lengthens the finishing period and the age of cattle at slaughter, increases sickness risk, and results in inconsistency of a quality gourmet product.

Cattle do not handle the unexpected well; domestic animals have been selected for routine processes, not to cope with hardship (survival) but with luxury (production in excess of survival). Stress increases not only under aggressive management during gathering or working in pens, but also due to lack of pasture and water supply, or during the every-day chores such as checking, moving to different paddocks, supplement feeding, etc.

Struggling for feed creates unrest, fighting and distress. Lack of feed and water are two main stress sources. Therefore, planning ahead for alternatives should allow you to minimize these stressors.

Adjustment of every-day management to a "minimum-stress" strategy is a simple and logical concept, but shifting from a fear-based to a curiosity- pleasure-based animal handling strategy is often difficult to implement.

In stocker/finishing programs especially, remember that steers are young and impatient in nature, with little or no previous experience in your corrals or around your electric fences and people. *To become routine, animals must be allowed to learn the routines first.* The time invested is worth the effort.

Colorado State animal behaviorist, Temple Grandin, said two things cause nervous, excitable animals. One was rough handling by humans and the other was genetics.

Ironically, while industrial agriculture has been designing facilities that put more stress on animals, they have been breeding animals that can handle less stress.

She said that animals that have been bred for leanness are much more excitable than animals that fatten easily. For example, commercial hogs and chickens today are now much more excitable than the older, fattier heritage breeds.

Genetic selection for single traits has deteriorated animals' feet and legs. She said lameness was now a major problem, and this added to the animals' stress as well.

A byproduct of this stress has been a decline in animal fertility and an increase in disease.

With early fattened cattle, she said the key to creating a calm animal was to always make sure the animal's first experience was a good one.

Allowing cattle to walk through a squeeze chute and be released with no pain prevents subsequent stress.

Grassfed cattle that are nearing slaughter can be taught to

not fear being hauled in a trailer by hauling them around periodically. She suggested that when you go to town to get an ice cream that you take your near-to-finish steers with you.

"That way when you load them to take them to the abattoir, they will think they are going for ice cream."

Animals that become upset need at least a half an hour to calm down before you can hope to work with them in a squeeze chute. Here are some other thoughts she shared:

■ Cattle like to be worked in small groups and not be crowded.

■ Never pet a ruminant animal on its forehead as this is a very aggressive act and an invitation to fight.

■ The way to create a bond with a young ruminant is to make it lift its head and stroke it up and down on the underline of its neck. This teaches the animal that it is submissive to you and that you are the dominant animal.

■ Ruminants also like to have their rumps scratched particularly above the tail head.

She said that graziers need to understand that cattle actually like new experiences as long as they have time to get used to them. For example, leaving the gate open to the corral will allow animals to become familiar with it at their own pace.

"Cattle really hate new things that are shoved in their face," she said.

Ann Wells is a veterinarian who works with ATTRA in Arkansas. ATTRA is the USDA extension outreach for sustainable agriculture.

"Disease is primarily caused by stress," she told me. Here are some other points she made:

■ An animal under stress can be visually identified by a body condition score that is sub-par to its pasture mates and by an excess of flies.

■ Flies are attracted to unhealthy, stressed animals.

■ Antibiotics are major stressors of animals and should only be used in very serious health situations.

■      Wormers are also animal stressors and should not be routinely used. Healthy animals with enough to eat shouldn't need to be wormed at all after seven months of age.

"The University of Tennessee has a large herd of cattle that have never been wormed and do fine," Wells said.

Replacement animals should be selected based upon their resilience to internal parasites and lack of flies.

She warned that cattle shipped to new environments are under stress. Consequently, cattle should be bought as close to home as possible and returned to pasture as quickly as possible.

It can require up to a year for a ruminant animal to adjust to a new environment and during that time it will be susceptible to disease. Many animals never adapt.

She said the use of preconditioning pens and unfamiliar feeds created a huge stress in pasture-raised animals and were a source of disease rather than a way of preventing it.

***Mud is the number one environmental stressor of ruminant animals.*** Cattle should be wintered on pasture rather than in muddy lots and barns.

She said that shipping cattle does not necessarily have to be stressful and that cattle learn from previous results. For example, cattle that have previously been shipped to better pasture readily load with little to no stress.

She knew of one grazier who routinely hauled stocker cattle to outlying pastures. All he had to do was to park the trailer and open the door and the cattle loaded themselves with no effort on his part.

New Zealand research found that most of the vaunted increase in tenderness had nothing to do with grain feeding and everything to do with the socialization of animals forced to be in close proximity to one another. Cattle that have been grazed at low stock densities are stressed when put together tightly for transport and/or at the abattoir holding pens. New Zealand researchers found that the occasional penning of cattle together prior to slaughter produced approximately the same increase in tenderness over range feeding as 150 days on grain.

American animal behaviorist, Bud Williams, said such socialization training doesn't have to wait until immediately before harvest but can and should be done long before. Williams said this can be done by just putting the animals in a pen together overnight. This allows them to get to know one another and feel part of a coherent herd.

Williams said cattle are herd animals and are actually less stressed when close to one another. He said they need to be reminded of that before you gather them for harvest.

Bud Williams said that if every time you put your cattle in a corral you hurt them with a shot, branding iron, or de-horner, you are creating a memory that being herded, put in a corral with humans nearby is a situation to be avoided at all costs. As a result, rounding them up will be stressful on both you and the cattle.

This shrinking of the flight zone is particularly important with cattle grazed at low stock densities and shifted infrequently. Range-land cattle, for example.

Cattle grazed in close groups and shifted daily, soon become extremely tame and are not afraid of humans and will quietly come to you to see if you are going to shift them to fresh pasture. If your animals mentally connect the presence of humans with fresh food, they will be very docile and unstressed around people even if they are total strangers.

Michel Hamel, president of FQRN Beef a branded beef abattoir in St. Lo, Normandy, told me that pre-harvest stress must be minimized for a tender beef product. Here are their rules for cattle handling at the abattoir:
- Never rush. Let the cattle take their time.
- Never hit them. (Electric cattle prods are not allowed with labeled beef.)
- No shipping longer than eight hours. (Three hours or less is best.)
- Slaughter within six hours after arrival. (Never longer than 24 hours.)

- Water is always available, including during transport.
- Soft music is played at all times in the holding pens.

## Fat Lowers Stress

The gamey meat syndrome I discussed earlier is also related to rate of gain prior to harvest. While genetics can help with tenderness, a high rate of gain prior to harvest is just as important as fat is necessary to lower meat pH.

Animals store energy in their muscles in the form of glycogen. Once the animal is dead, the muscle glycogen converts to lactic acid. This causes the pH of the carcass to fall from the live animal pH level of 7.1.

The more glycogen available, the more lactic acid will be produced, enabling the pH to fall to the acceptable range of 5.7 or below.

If there is not enough glycogen available, insufficient lactic acid will be produced. This will result in the pH level remaining high, causing dark cutting, toughness and off-flavors.

In other words, *a well marbled animal can handle stress* — as it relates to meat quality — *much better than an animal with no marbling.*

To put this in scientific terms, a pH of 5.5 requires a muscle glycogen content at slaughter of at least 57 moles per gram. The greater the amount of muscle fat above 57 moles the more of a buffer zone the meat has against stress.

Meat pH also affects meat color. The closer the meat pH is to 5.4 the brighter red the meat will appear. This is why the Japanese grade beef primarily on its color.

Australian research has found that for an animal to have sufficient glycogen reserves in its muscles to drop the pH it must gain over 1.8 lbs per day for a minimum of 30 days prior to harvest.

This is why gamey meat tends to be more of a problem with late summer and fall harvested grassfed meat than spring and early summer harvested animals. There are no perennial grasses capable of producing late summer and early fall gains high enough for intra-muscle fat to quickly drop the meat pH.

Because feedlot gains tend to always be in excess of two pounds a day prior to harvest, this gamey flavor problem is not nearly as big a problem with grainfed beef.

### Time in Transport Is Important

Keep in mind that animals start burning fat the minute they are removed from the pasture. Argentine research has shown that an animal can live off the energy stored in its liver for eight hours. After that it will start to mobilize the marbling fat stored within the muscle. This marbling fat will be totally consumed within 24 hours of being off feed.

While time in transport is very important, research by Dr.

---

**How to Identify Stressed Cattle**

■    Stressed cattle stand uneasy and will not turn their backs unless running.

☐    Unstressed cattle are curious and eventually walk away to graze or lay down to ruminate.

■    Stressed cows look and call for their calves.

☐    Unstressed cows graze and are careless about the location of the offspring.

■    Stressed cattle eat grass fast, with small bites.

☐    Unstressed cattle harvest grass with large and slow bites, with time devoted to mastication.

■    Stressed cattle forced into water drink only once.

☐    Unstressed cattle wait for access to the water point and drink more than once before leaving.

■    People around stressed cattle complain about electric fences.

☐    People around unstressed cattle believe electric fences are a key management tool.

**Dr. Anibal Pordomingo**

Burt Smith at the University of Hawaii found that transport itself was not stressful as long as the animals were with familiar herdmates. In fact, he said they appeared to enjoy it.

Animals develop social groups where every animal knows its place. If you bring strange cattle into a group everyone has to refight for their new status and this expends a lot of energy. Consequently, *finishing cattle should never have new cattle introduced in the 14 days prior to harvest.*

Also, make sure your abattoir has separate holding pens for cattle awaiting harvest so that your cattle aren't mixed with someone else's.

Prey animals, like cattle, protect themselves from danger by herding in a tight knot. Predators will not attack such a tight group for fear of getting hurt, so the herd is a safe place to be, and the presence of other animals is very reassuring.

However, if you take a prey animal and put it in a novel situation by itself, you have an extremely emotionally stressed animal.

This becomes a real problem for graziers who want to take a single animal to the abattoir. One solution to this is the use of a "Judas Goat."

This is a companion animal that rides to the abattoir with the animal and stays with it until it is harvested. Female goats are an excellent choice for this valuable service and have been used for centuries.

### No Fear of Death

Many of us assume the animal is stressed by the sounds and smell of an abattoir. However, there is little evidence of this.

Humans are the only animals that have the knowledge that they will die. All other animals interpret the future by what has happened to them in the past. If what is happening now is similar to what has happened before with a good result, they are not stressed by it.

As my Dad put it, "Animals are happiest when today is just

like yesterday."

Since no animal has a memory of death, animals do not fear it or feel threatened when they witness it in a companion.

Many of you have probably seen a male dog run over by a vehicle in the presence of its running pack. Did you notice that none of the other dogs who witnessed the death were particularly upset by this event, nor did it stop them from chasing cars?

Graziers who field-harvest bison with a rifle say the targeted animal's herdmates hardly flinch when one falls.

However, because animals cannot interpolate a future, ***they are stressed by novelty.*** Strange sounds, strange people and unfamiliar cattle and surroundings start them burning muscle glycogen like gasoline.

While it is often said that animals are stressed by loud sounds, they are stressed by loud unusual sounds.

Cattle who grow up next to a railroad track are not stressed by the loud passing of the train and in fact seldom look up as it passes.

French abattoir protocol specifies the playing of music in the holding pen area. This is to provide a "white noise" to mask sudden, unusual sounds that might stress the cattle.

At FQRN abattoir in St. Lo, Normandy, the music of choice is Classical. I asked abattoir president Michel Hamel, if this music choice was to calm the cattle.

He replied, "No, it is to calm the workers."

Think about the noise tolerance of a feedlot animal compared to a pastured one. The feedlot animal lives in a world of growling feed trucks and grinding feedmills and soon becomes habituated to them.

Now, put all of what happens to an animal at a feedlot — the close confinement to its herdmates, the frequent presence of people, noise and vehicles — together with an average daily gain of at least two pounds and you suddenly realize why gaminess is seldom a problem in feedlot animals.

Anibal Pordomingo suggests the following check list for lowering stress both on the ranch and at the abattoir.

**Do:**
- Run on flexible time schedules.
- Make sure you know what to do and how before walking into the pens.
- Make sure the crew knows how you want cattle handled and remind them periodically.
- Use as few people as possible. Three can deal with sorting and moving up to 800 to 1000 animals in corrals with ease.
- Work on foot and follow routine strategies. Cattle will remember them.
- Use visible light sticks to point directions, and touch cattle to make them notice your presence but do not hit them.
- Make yourself and your help visible to the animals.
- Expose all your tools and sounds to the cattle before starting to work them.
- Lead as much as possible.
- Let them recognize your presence, not your aggressiveness.
- Let them have time to perceive the layout of the place and alternative paths before invading (reducing) their vital space.
- Water and hay in holding pens can help to relieve tension. Thirsty cattle are always stressed.
- Make working, moving or sorting cattle an easy, no-big-deal job.

**Don't:**
- Do not plan on time-short inflexible schedules.
- Do not use untrained and excess help.
- Do not have "help" who do not know what to help with.
- Do not make handling cattle a rodeo opportunity.
- Use no dogs at any time.

- Do not use whips.
- Do not yell or make unexpected noises.
- Do not create surprises for the cattle.
- Use no electric hotshots.
- Do not use horses in corrals unless absolutely necessary.
- Do not provide escape routes for the cattle.

# CHAPTER 12

# Turning Cull Cows
# Into Gourmet Products

There is probably more profit in your culled cows than in your steers if you do it right. However, there is marketing peril if you do it wrong.

The problem is that most North American beef producers believe that the meat grinder solves all meat tenderness issues. In fact, it is looked at as sort of a garbage disposal for animals too thin, too lean, or too tough for "table grade" steak.

In contrast, the French warn that a "gourmet" quality ground meat requires a more tender animal than a steak because there is less you can do with cooking time to manipulate its tenderness. However, this does not mean that a culled cow cannot be a "gourmet" eating experience and that it cannot be served up in higher value cuts than just hamburger.

Paris native, Jerome Chateau, said the wide-spread American belief that meat from older animals has to be tough strikes most Frenchmen as incredibly naive. In fact, given the choice — as they are — the extremely picky French actually prefer their beef to be from older animals.

While the French have a large grain-based feedlot system that produces some 31% of the country's total beef production, only 11% of the country's internal beef consumption comes from animals less than two years of age and only two percent is from young males (mostly as dairy veal). Eighty seven percent of internally consumed beef is grass finished and

75% of French beef consumption is from culled cows.

Chateau said that French dairymen will typically voluntarily cull 30% of their cow herd each year for beef sales.

Cows that are to be sold for beef are allowed to graze and fatten for at least 90 days after being dried off. He said a typical practice was to dry the cows off in the winter and sell them as beef the following summer after they had fattened on the lush spring grass.

In other words, their cull cow protocol is the same as for "table grade" beef.

Since the outbreaks of BSE (Mad Cow Disease), labeled beef has become a big item in France. Many upscale French supermarkets no longer stock unbranded commodity beef. Currently, 42% of the beef eaten in Paris is labeled.

Chateau said the oldest labeled beef brand is the Normande breed label. This highly successful beef label is 95% from culled dairy cows. He said the average age of the cows used for the premium priced beef was six years.

The Normande beef label requires:

■ The cattle have to have been on one farm for at least four months before slaughter.

■ The cattle must spend at least eight months of the year on pasture.

■ The cattle must not spend over four hours on the truck to the abattoir.

■ The cattle must not stand for over 24 hours at the abattoir before slaughter.

■ The cattle are to be kept in individual stalls with free-choice water.

■ The cattle must be handled with a minimum of disruption.

■ The beef must be aged for seven days as a carcass or for 12 days in plastic wrap as individual pieces.

Chateau said that as French awareness of the benefits of CLA and Omega-3 has grown the beef labels have tended to require more of the animal's year to be spent on pasture. He expects the beef labels to become even more restrictive in the future.

In Paris, I visited with the proprietor of the Boucherie du Square Voltaire. He showed me that each piece of meat had a label on it that allowed it to be traced back to the farm it came from. He even had a picture on the counter of the steer he was selling the meat from that day.

He said ***the best beef in France was produced in May and June*** when the pastures were at their peak. What was very interesting was their beef pricing.

The highest priced beef was from a 48-month-old steer.

The second highest was from a 30-month-old heiferette.

The third highest was from a cull cow five to nine years in age.

The fourth highest was from a steer older than 30 months of age.

Meat from male calves 500 to 800 lbs was sold as veal and brought the lowest price.

The meat cutter said he considered the best flavored meat to be from a five- to nine-year-old cow. The older cows marble easily and are considered by the French to be in the prime of life.

"A five-year-old cow is like a 36-year-old woman. She is at the peak of her beauty," he said.

The French dairy breeds are all dual-purpose breeds. This means they are both milked and produced for beef. Unlike the Jersey and the Holstein, which have poor muscling, the French dairy cows have a "beefy" look to them because beef production is as important as milk. In fact, the milk provides the cash flow to grow the cows to slaughter.

Most dual-purpose breed males go for veal. There is even a special Label Rouge brand for naturally suckled veal. The Alpine Aubrac breed cows are milked in the morning and the calf is allowed to suckle in the evening.

At the Paris International Agricultural Show, I found it interesting that the breed associations offered visiting consumers samples of beef and cheese of their breeds. Most North Americans never consider that each bovine breed has its own unique flavor. Surprisingly, the best tasting beef I sampled was from a nine-year-old Aubrac cow!

The French appear to be onto something with their belief that flavor is correlated with age. The absolutely most memorable meat I have ever eaten came from a 20-year-old bison cow in Hawaii.

There is an all-you-can-eat buffet near my home that serves pre-cooked steaks cut from local cows. These yellow-fat steaks are far more flavorful than those sold for a small fortune at the "Prime" steak house in our town. Because they are slow cooked they are just as tender as the grilled steaks at the steakhouse. We also have a world-famous barbecue restaurant in our town and culled cow brisket is the only beef they use.

Similarly, most of the "country fried steak" you eat in restaurants is from culled cows. Breading the meat hides the dimpled look of the Jacard tenderizer. Even the tenderloin served in a "Prime" steakhouse is just as likely to have come from a culled cow as a steer. The point I am trying to make is that there is much more that you can do with a culled cow than just turning it into hamburger.

Now, *there is a tradeoff to this increase in flavor in cow beef and that is in texture.* Beef from older animals does not have the fine grain of beef from younger animals. This is often described as "stringiness" and some people object to this. It doesn't bother me but you need to be aware of this potential problem.

Also, there is a lingering distrust of yellow fat, although this is dwindling as more consumers realize this "yellow" is carotene — a precursor of healthful Vitamin A.

In a Brussels, Belgium, meat shop I saw that beef with yellow fat was priced higher than that with white fat. I asked the Belgian chef I was traveling with why this was so.

"The yellow fat is the ultimate proof it is a grassfed animal," he explained. "The yellow fat is very good for your health."

The other tradeoff is that a cow whose final destination is as a "gourmet" meat product must not be starved at any time in her life. The tradition of "wintering cows rough" won't make a gourmet meat product.

And, this is true even for ground meat. Or maybe, as the French believe, particularly for ground beef.

The New Zealanders have built a highly profitable industry on growing out two-year-old dairy bulls for American hamburger chains. This extremely lean but tender meat is precision mixed with the fat trimmings from domestic grainfed beeves in a formulaic fat-to-lean ratio designed to insure that every hamburger you eat at that restaurant will taste exactly like the last one. Nothing is left to chance.

I have noticed that some grassfed direct marketers have no problem selling hamburger and others have great difficulty. I think the key element here is that the ones having problems *have not appreciated the necessity for "finishing" their culled cows* and possibly selling this meat separately from the ground meat from their two-year-old steers.

James Girt, owner of River Run Beef in Clatskanie, Oregon, said he quickly recognized the eating difference between his culled cows and his grass-finished steers. Girt has built a premium priced niche for his "gourmet" organic ground beef with Portland "gourmet" pizza restaurants.

"The ground beef from a two-year-old, grass-finished steer is entirely different than that of a culled cow," he said. "We prefer to send our culled cows to the auction rather than develop a separate market for it."

In contrast there is a grassfed producer in Colorado who winters his cows "off the fat on their back" on desert range. A

---

**You may think you are sending that old, thin, broken-down, cow "to MacDonald's," but you aren't. MacDonald's has stringent quality control protocols to protect the reputation of their ground beef.**

consumer described the ground meat product he had purchased there as similar to "eating BBs."

Needless to say this producer has a constant problem moving even the quality ground beef from his finished cattle because consumers who have gotten a package of BBs aren't willing to risk buying from that ranch again.

***The meat grinder is not magic. It cannot turn tough meat tender***. If you don't have the grass to properly winter your cows during their life and to "finish" your culled cows before harvest, you will be a lot better off sending them, as James Girt does, to the auction or, possibly, to the dogs.

Sending too-tough grassfed beef to the dogs as dog food may not be as unprofitable as you might think. Particularly, if it is a "gourmet" product.

Ozark grazier, Kendrick Ketchum, has developed a premium-priced dog food market in the Little Rock, Arkansas area.

Ketchum who raises grassfed beef and pastured poultry in Heber Springs, Arkansas, was approached by a wealthy Little Rock attorney who was an animal rights activist. She had read about the health benefits of pastured meat products and had a business proposition for him.

She wanted a custom-made, 30% grassfed raw meat and 70% fresh vegetable produce dog food made for an abused animal shelter in Little Rock. She was willing to pay the going retail rate for the meat and vegetables to get what she wanted and wanted beef, lamb and chicken versions. The chickens were to be ground whole including the bones.

The ground meat was mixed with ground up zucchini, okra, yellow squash and cucumber from his garden, minerals and kelp, shaped into patties and freeze-dried. He said this was an extremely easy product to make because everything went through the meat grinder.

The price Ketchum got for this concoction was three dollars a pound or roughly what he was getting for a grass-finished beef carcass.

The other "hot" doggie product Ketchum cooked up —
or rather smoked up — was smoked beef bones for dogs.

Using the large leg bones from his beeves and putting them
in his abattoir's smoker, he has been able to get $5.00 a piece for
the six-inch bones, which come complete with a red ribbon.

Smaller bones are packaged in a clear plastic bag and sold
for 75 cents a pound in five pound packages. He said such boney
creativity could add $100 or more to the gross of each grassfed
steer.

# CHAPTER 13

# Abattoirs

Abattoirs, which are perhaps more correctly termed animal disassembly plants, can be profitable over a wide range of volume, even down to just a few head a day. However, no size can be profitable without a reliable daily throughput of cattle.

*The hidden cost of seasonal production is that it will not allow the building of the semi-industrial, small volume abattoirs most graziers would like near their operations. Note, I said semi-industrial.*

Industrial scale kill-plants can use a very low-grade labor by breaking the disassembly process down into simple repetitive cuts that can be taught in a few minutes. With employees in the hundreds or thousands, these plants require a huge amount of daily cattle to break even but can kill, process and package an animal for around $50. Such large-scale plants were not possible until the rise of the huge custom feedyards in the 1970s, which greatly concentrated the supply of cattle.

In business, supply creates demand. It is the existence of cattle that creates abattoirs, not the reverse.

So, how do you start something new with low volumes of production?

As Teddy Roosevelt said to his men pinned down by enemy fire in Cuba, "Do what you can, with what you've got, where you are."

The North American grassfed industry today is primarily made up of small producers selling frozen meat through farmers' markets. This is the case because this sales and distribution method is simple, direct and produces a large financial return per head sold.

While large-scale producers normally look at return on investment as a judge of profitability and seek out markets capable of absorbing huge investments, small-scale producers look at the per head return because most do not see large scale expansion as economically feasible or desirable.

The consumers who buy at farmers' markets not only are willing to pay often exorbitant prices but are also more tolerant of the uneven meat quality that typifies much of today's grassfed product. All other markets for grassfed beef offer a lower return per head than the direct sale, demand an even quality and a fresh product, and so are not as attractive as the direct sale to low volume producers.

*I like to think of the farmers' market as the lowest hanging fruit on the tree.* It is easy to pick but sooner or later you are going to need a ladder to climb above the competition.

All small volume pioneering efforts have to target premium-price niches because their costs are much higher.

---

**Unlike many grassfed producers, the Gamble family of St. Helena, California, sell fresh beef rather than frozen. This fresh meat policy gives them entree to some of Napa Valley's top restaurants.**

**Restaurants do not like to buy frozen beef because the meat "bleeds" on the plate due to ruptured cell walls. Many customers find this bloody juice unappetizing in appearance.**

**In 2005 they took two animals to the abattoir every week, 52 weeks of the year. Thanks to this steady week-to-week business, he has a good relationship with his abattoir.**

While it is true the feed costs of a grassfed product are lower, the overhead and abattoir costs are much higher due to the current low volume.

The point I am making here is that at this time a new whiz-bang, large-scale slaughter plant would not dramatically increase the amount of grassfed cattle because the current participants are not interested in large-scale production. The importance of producer attitudes is often lost on abattoir consultants peddling pre-written business plans. New things don't just quickly "scale up."

So, where are we today?

With the exception of the West, there are few areas in rural America that don't have a small custom abattoir within a few hours drive of virtually every grazier. Your State Department of Agriculture will have a list of these in your state if you are interested.

*Small volume producers and small volume abattoirs are a part of the same production whole.* However, small volume abattoirs require a much higher grade of labor than factory kill-plants because the labor must know more to do more of the job. This makes their labor costs per slice of the knife much higher than the industrial-scale plant.

Small abattoirs have a hard time attracting part-time labor even in areas of high unemployment. Consequently, they have to retain their higher paid labor full-time even when there is no work for them to do. It is this inactivity that really drives up labor costs.

These higher labor costs mean a small volume plant can never compete on a head-to-head basis with a large plant. To survive, they must find a niche consumer willing to pay a premium price for their services.

Many pioneer grassfed producers are currently paying as much as $300 a head for kill, fab and wrap or five times the industrial cost.

However, this rate is very negotiable if you can offer a consistent level of volume or target his slow season. In most areas, this will be the spring and summer as processing game typically occupies the fall and early winter.

All abattoir owners are looking for a customer who can provide a base level of business to cover their overhead "nut." They can then make their profit on custom kills, sale barn cows and spent bulls.

Abattoir owners tell me that a plant with its base overhead covered can be a very, very profitable business doing custom work.

### Prototype for a Small Abattoir

The Remer Meat Company in Clinton, Missouri, has been touted by some as a prototype worth studying for a small-scale abattoir and meat processing plant.

The 9000-square foot building includes a livestock holding area, slaughter room, cooking facilities, smoke house, processing area, an office for the USDA meat inspector and retail store.

The plant can slaughter and process beef, pigs, sheep, elk and bison. It has a maximum kill capacity of 25 to 30 beeves a day and refrigerator space for 100 carcasses.

The plant can make beef jerky, summer sausage, cubes, cutlets, patties and has a 400-lb capacity oven for smoking and curing meat. It also has a Jicard mechanical meat tenderizer for the tougher cuts. All of these optional services are charged in addition to the slaughter and processing fee.

Owned and managed by David and Sue Remer, the plant has eight employees. It currently only kills on Tuesdays and Fridays. If run on a five-day kill schedule the plant would require approximately 20 employees.

"Finding good employees is the toughest job," David said. "In a small custom plant you need flexible people who can do multiple jobs."

The plant is profitable and brings in about $40,000 a month in monthly cash flow. Sixty percent of this is from retail meat sales and 40% from custom slaughter charges. David

Remer said these figures are exactly opposite of what he would like to see.

"The profit in a slaughter plant is in killing animals — not in retailing meet," he said.

David said ***an ideal customer for him would be a slaughter customer who could guarantee a certain number of animals each week.***

"An abattoir needs a base level of steady business to be profitable. Without it, custom slaughtering is a crap shoot every day," he said.

Remer Meat currently kills and processes beeves for the Ozark Belgian Blue Cooperative and would love to have more such steady customers.

The plant charges a $28-a-head slaughter fee and 34 cents a carcass pound for processing. However, David made it clear that for a volume customer "everything is negotiable."

If you are only going to market a dozen head a year, you are not going to have much leverage for negotiation, and can, in fact, find yourself shut out by bigger, more reliable customers.

Ridgeway Shinn of Hardwick Beef in Hardwick, Massachusetts, said the extremely popular custom abattoir he uses will not allow him to increase his contracted weekly kill by even so much as one head during the busy fall deer processing season.

The problem with relying exclusively on one small abattoir is that you can get in the situation where ***the shoe dictates how large your foot can grow.***

If you are a growing business, eventually this is going to pinch.

So, what about doing it yourself?

Shinn said to not even consider this if you do not have a steady year around output of cattle. He has done this with his New England business and said while not easy, it is do-able.

The other key factor is — are the people who run the plant part of the deal? He said there are lots of closed abattoirs for sale but without the highly trained people such small operations require they are worthless.

One alternative is learning butchering skills yourself.

If you do have a year-around supply, he said to consider that there are really two functions involved in an abattoir. One is the killing of the animal and the second is the fabrication of the meat.

It is in the fabrication portion where all the value is added. Many custom abattoirs will kill the animal and skin it for $50 a head if they can keep the hide.

"The art (cost) in meat-cutting is in the fabrication, not the killing."

He thought the ultimate for a grassfed branded beef company would be a centralized fabrication facility with portable abattoirs feeding carcasses to it from on-farm slaughter.

He said getting a zoning permit for a new kill floor was a real hassle as people imagine that it will be smelly and attract flies. There are seldom zoning problems with an on-farm fabrication facility. Just call it a butcher shop.

## Do It Yourself

Jon and Wendy Taggart of Burgundy Beef in Grandview, Texas, use a nearby custom kill plant but do their own meat fabrication.

Taggart, described his relationship with his previous custom abattoir as a "nightmare."

"I couldn't ask for anything the least bit unusual. He wouldn't vacuum package the meat and didn't want to age it. I was spending $30,000 a year with the man and still couldn't get any respect.

"I finally decided I couldn't beat the packers, so I became one."

In September of 2004, he and his wife, Wendy, opened their brand new 2800-square foot combination USDA-inspected, meat fabrication and retail location on the northern outskirts of Grandview ( population 1200), located 35 miles south of Fort Worth on I-35W and near their ranch.

The butcher shop is called "The Burgundy Boucherie. A pastured beef butcher market and purveyor of fine artisanal foods."

Jon said the company name clearly reflects Wendy's marketing influence. While the French name gets them a lot of ribbing in Grandview, it has been a hit with their sophisticated North Dallas customer base.

"Wendy and I have a clear division of labor," Jon said.

"I am responsible for the animals while they are alive and she is responsible for them once they are dead."

Wendy really enjoys having a face-to-face relationship with their customers and said she would never want to just wholesale meat.

"By being able to talk to them in person, I can tell them about ways to cook cuts of meat they would never order other-wise," she said.

The Taggarts have built their business on delivering highly marbled, grassfed meat to upscale Dallas customers. They have a minimum order of 10 pounds and their average order is 20 pounds or around $100.

A popular seller is a 100 lb variety box.

When I visited, they were completely sold out of ground beef. "That's a real high class worry to have," Jon said with a laugh.

Wendy said many of their buyers were stay-at-home mothers who were home-schooling their children. Luckily, these Moms buy a lot of ground beef and selling all the ground beef they can produce has never been a problem.

"Our dry aging makes our hamburger sell itself," Jon said.

The frozen meat is ordered by e-mail and delivered once a week direct to the consumer in the metro area. Every Monday morning Wendy e-mails her customers and tells them she will be coming to Dallas on Wednesday and that there is still time for any special-order cuts they might like. "Standing rib roast is a very popular holiday special-order cut," she said.

The Taggarts have a 90% re-order rate and most advertising is by word-of-mouth. However, Dallas radio personality, Howard Garrett, has been a big help with frequent plugs on his Sunday morning organic gardening show.

"Our typical customer is a well-educated and nutritionally aware North Dallas mother," Wendy said.

Food buying co-ops are another large customer group they serve in Dallas.

Jon said they started selling halves and quarters but *selling by the cut is much more profitable.*

At the Grandview store they sell their own beef and buy lamb, pastured poultry and pastured dairy products from other graziers in their area for resale. In their first month after opening their new store, retail sales doubled.

Jon attributes this primarily to having enough room to add lamb and poultry to their product list. Surprisingly, in-store fresh meat sales have been slowly building from Grandview customers as well. "We never thought we would attract rural customers to our store but we have," Jon said.

A key component Wendy insisted on for the facility was a demonstration kitchen. They also plan to add an outdoor Argentine-style parilla grill.

"Teaching people how to cook meat in new and different ways is the major way I sell our meat," Wendy said.

She plans to invite Dallas area garden clubs to Grandview for a "day in the country" where she will demonstrate a lot of her European-style meat dishes.

To ease potential zoning controversies, the Taggarts decided not to do on-site slaughter and use a USDA inspected abattoir located 30 miles away.

The custom killer charges $40 per head, skins the animal and cuts the carcass into quarters. The quarters are then transported to the Taggarts' fabrication facility where it is aged and then processed.

All of the Taggart's meat is dry aged for 21 days before selling.

Due to cooler space constraints at their previous abattoir, the Taggarts had been forced to go to a weekly harvest even though their meat was sold as frozen.

Jon said this lent itself well to adding their own facility as a constant year around supply of animals is critical to making an abattoir investment work.

Taggart custom grazes 1200 to 1500 head of yearling cattle for a Fort Worth investor. This provides both a stable monthly cash flow and reduces the amount of capital he has to have tied up in cattle inventory.

He has worked a deal with the investor whereby he can select cattle for grass finishing out of this large group by paying a premium of $100 a head over the feeder cattle market.

He uses Angus heifers exclusively for his grassfed program.

---

**One August, I went to Argentina to check out the effect their winter had on the beef supply and quality. I asked Horacio Bravo, meat quality control officer of the CEBA packing company in Pontevedra if there was a seasonal decline in the number of finished cattle in the winter. CEBA is a major export abattoir known worldwide for its high meat quality.**

**"No," he said.**

**"Why not?"**

**"We raise the price."**

**In the West where ranches specialize in finishing, there is little seasonality in production. Bravo said most ranchers liked to sell 35 to 40 head weekly for cash flow purposes. (This is a truckload in Argentina.) As quality control officer he would like every animal killed to be an Angus weighing 1000 lbs with a good fat cover (but not excessive), 24 months of age.**

**Brave said Argentine beef has a shelf-life of 120 days and is 90% consumed by 60 days of sale.**

"We have built our reputation on marbled meat and absolutely refuse to harvest an animal that is not well-fattened."

Unfortunately with today's super-sized Angus, this frequently means taking the heifers to 1300 to 1400 lbs before they are properly fattened.

"I have found you can farm out everything in grassfed beef but the actual grass finishing. There are just too few people who know how to do this well."

To keep stress at a minimum, *Jon uses a lead steer and the heifers are penned frequently in the corral* and then released to "pattern" them to being comfortable with the pre-harvest roundup and transport. Cattle selected for harvest are always kept with a known companion at the abattoir.

"Our ultimate quality control is that we eat one steak out of every animal we harvest." Jon said.

He said if you have developed a forage chain capable of finishing cattle year around, adding a personal abattoir is something to seriously think about, but to scale it small at first.

He received a lot of encouragement from Minnie Lou Bradley, a Texas specialty meat marketer, but found Texas A&M to be overwhelmingly negative about grassfed meat in particular and direct marketing in general.

However, despite the Extension Service's misgivings, they decided to proceed with their plan and found their timing was excellent.

"There has never been a better time to be buying meat cutting equipment," Jon said. "Many of the supermarkets are closing their in-store meat processing facilities and you can buy this equipment today for pennies on the dollar."

---

**Suspending the carcass by the aitch bone rather than the Achilles tendon stretches the muscles of the round and loin to increase sarcomere length and increase tenderness.**

**Texas A&M**

These bargains can often be found on the Internet. He purchased his meat grinder and scale on Ebay.

He estimates his total investment in the dual-purpose facility at $200,000 but said he did much of the interior finish work himself.

"We were previously netting between $700 and $1000 a head but we expect to do better than that with our own plant due to our ability to do more added-value processing."

Wendy learned the basics of meat cutting by helping process their meat in the previous custom abattoir.

Currently, the Taggarts are processing four to five beeves a week from their own ranch utilizing three part-time meat cutters.

However, *the facility was purposely built oversized* to allow for easy expansion and could handle three times the current number of animals per week with no increase in capital costs.

This over-build has allowed them to do custom-processing for other Southwestern grassfed producers and they welcome this business. The facility was built so that it can easily be doubled in size and Jon estimates that such an expansion will be necessary.

An urban site was chosen rather than an on-farm one because they had found that urban consumers are squeamish about seeing live animals when they are buying meat.

Jon said they have gotten a lot of customers from Jo Robinson's website www.eatwild.com including one very persistent Fort Worth restaurant that purchases all of its menu items locally.

"We explained that we were not interested in wholesaling but he was insistent that he had to have our beef. We finally compromised by selling him ground beef wholesale and 35 high-end cuts a month at our regular retail price for his special events."

Another area chef buys their marrow bones and tails.

"Luckily, due to Texas' barbecue tradition, we have a

waiting list for brisket," Jon said.

Jon finds direct marketing and custom grazing far less stressful than the commodity stocker business he once was in.

"I don't even pay attention to commodity cattle prices anymore and have developed a high-margin, low-volume philosophy. *We don't need nearly as many cattle to make a good living as we did in the stocker business.*"

## On-farm Abattoir

Nathan Creswick's 300-acre farm is near Grand Rapids, Michigan, and is located near the main freeway between Detroit and Chicago. The farm has its own on-farm abattoir.

The 1800-square foot abattoir is operated as a stand-alone custom slaughter house and does custom work in addition to the farm's grassfed beef and pastured pork.

The abattoir does nothing but process deer during hunting season.

"Hunters that don't get a deer always buy a side of beef, so we root for the deer," Creswick said.

The abattoir operates under the name of Pioneer Beef Processing. It is State inspected but is built to USDA specifications if Federal inspection is desired later. The abattoir is managed by Creswick's nephew, Paul Suits.

"We have it designed so the cattle and pigs walk right into the abattoir off the pasture. This keeps transportation stress off of them."

All washdown water from the abattoir is sprayed on the farm's compost pile. This is a practice the State inspectors really like.

The abattoir has space to hang 40 beef carcasses for aging. All grass-finished beef is aged for three weeks before processing and freezing. All beef and pork is sold frozen.

The State inspection of the on-farm abattoir means that all beef and pork going out of state must be processed by a USDA inspected plant. Luckily, a USDA pork abattoir is only three minutes away and a beef abattoir 45 minutes.

The farm is currently harvesting three to five beeves a week and four to six pigs. Grass-finished cattle from Alabama are imported to even out the year-around supply.

Health issues are important to Creswick's customers.

"Currently, 30% of our customers are people who have cancer or who are recovering from cancer," he said.

"We also have a lot of former vegetarians who have been attracted by the humaneness of our production and harvest techniques. And, of course, we have a lot of Atkins diet followers."

Creswick said his current marketing mix is one-fourth to restaurants and the remainder sold on-the-farm or to consumer food buying clubs in Detroit and Chicago.

The Chicago buying club consumes about four to five beeves a month and the Detroit club consumes that amount every three weeks.

## Diner Finds Profits in Processing

It has been said that profit is found in the problems. Tod Murphy of the Farmers Diner in Barre, Vermont, would have to agree.

Murphy's idea was to recreate the idea of "community" food with a small 60-seat diner that would serve food grown by local farmers. A grass farmer himself with a small sheep dairy, he said he thought it was important that consumers have a connection to their food.

He purposely chose a diner rather than a white table cloth restaurant because regular people could relate to it. He wanted Farmers Diner to recreate what he describes as "an American icon."

Barre is a small town of only 12,000 people but Murphy has found it to be a cheap and forgiving place to make the mistakes common to all start-ups.

Obtaining locally grown vegetables, eggs and dairy products was no problem. The problem was in obtaining locally produced meats.

"The first thing I learned about trying to source local food is, meat is the issue," he said.

More specifically the problem was that there was no meat processing infrastructure left in his area. While this would have sent most restaurant entrepreneurs to the phone to call IBP, Murphy refused to give up his dream and decided to build his own.

The corrugated metal building he built was described by the *New York Times Magazine* as "one of the smallest USDA inspected meat-processing facilities in the country."

Here five employees cut steaks and chops and smoke and cure bacon, ham, sausage, turkey, fish and cheese. Murphy said the plant set him back $160,000 and is not licensed to kill.

*Due to water quality protections, slaughter plants are much more expensive to build than processing plants.* Murphy has his animals harvested on a custom basis by a USDA-inspected abattoir. This is no great loss as *the majority of the value is added in the processing.*

While built as a necessary evil, Tom has discovered the meat processing plant to be a source of "unfair" competitive advantage for his restaurant.

"There is real money in meat processing," he said.

While most meat marketers have problems marketing their hamburger and lower end pork cuts, those were the very items his diner most heavily used.

---

**Yield Makes A Big Difference**
**Carcass yield is directly related to carcass muscling percentage, degree of marbling and the muscle to bone ratio. On an 1100-lb steer selling for $3.00 a carcass pound, a one percent increase in yield adds $33 in profit. Now, consider that 10% differences in carcass yield are common. That's a whopping $330 a head difference! In direct marketing, superior genetic selection can pay big dividends fast.**

"For example, I primarily want the bacon and sausage from a pig. I can then sell our whey-fed tenderloin and chops to white table cloth restaurants for a premium price. The chefs just love the taste of whey-fed pork.

"Similarly, the way to get the best price on a T-bone steak is for the buyer to take a whole steer. *The reason the prime cuts are so expensive is because most processors have to give the hamburger away.* That's not a problem for me."

Having found a source of unfair advantage, Murphy has devised a strategy to maximize it in what he calls his "pod" approach to future expansion.

Recognizing the processing plant as the profit generating centerpiece, each "pod" can supply up to eight Farmers Diners within a geographic region.

These processing pods would also produce sausage, smoked cheese and tomato sauce under the Farmers Diner label.

His current tiny processing plant can supply three restaurants and he plans to open a larger 150-seat restaurant in Burlington, Vermont.

"A 150-seat diner consumes 2.5 beeves and 35 hogs a week. When you are paying a premium price for your food the only way to cut costs is through increased volume."

He said restaurant sales do not follow seating additions on an incremental basis.

"A 60-seat diner can do $450,000 a year. A 150-seat diner will do $2.5 to $3 million.

"You have to have the capacity to maximize peak times. It's all a game of numbers and percentages."

Murphy currently has 35 to 40 farmers who supply him with 70% of the food he serves. Some of this, such as his whey-fed pork, is contract production. He foresees the percentage of local food rising to 80% in the future.

The Diner is priced at a "modest premium" over competing diners and attracts an eclectic mix of customers.

## Portable Abattoir Could Have National Impact

Until recently, all livestock in the San Juan islands of Washington State had to come and go by ferry. Unfortunately, this frequently resulted in manure spills on the ferry deck. Clean water regulations prohibit washing the manure into the Sound, and this created a major problem for the ferry companies. As a result, the island community became interested in developing a way that the animals could live and die on the islands without having to be transported to the mainland.

Initially, a stationary abattoir was proposed on San Juan Island. Even though this was to be located on government-owned land near the island airport, the proposal created a furor among island residents who feared strong smells and flies.

An alternative idea was a small portable abattoir that would do the actual harvest on the farm. In other words, *the abattoir would come to the livestock rather than the reverse.*

This idea found no opposition from island residents but, of course, there was no precedent nor a known manufacturer of such an abattoir. There was also a lot of skepticism whether such an abattoir if built would meet USDA standards. This has proven to be unfounded.

Intrigued by the economic development possibilities of such a plant in other rural areas, the USDA provided a Value-added Grant to fund the research and testing of the portable abattoir idea.

Co-op secretary, Bruce Dunlop, took on the point man's job and journeyed to Texas to look at a portable abattoir being used to harvest domesticated deer. From what he saw and learned on that trip, Dunlop sketched out a plan for a small portable abattoir that could be used to harvest cattle, sheep or hogs.

The Lopez Community Land Trust — a non-profit group that promotes sustainable rural development in the San Juans — agreed to pay for the prototype to be built and to then lease it to the farm co-op. This secondary funding source was necessary because the USDA grant could not be used for capital expenditures — only for design and research.

The abattoir was built by a Washington trailer manufacturer to a gooseneck design and can be pulled by a one-ton truck. The gooseneck coupling allows for both easy trailing and backing, which is important as the unit often has to be backed onto the ferries.

The need for a "ferry fit" also limited both the height and length of the unit making it somewhat smaller than ideal. Since ferries charge by the linear foot for carriage, the abattoir was made only 13 feet long.

Including trailer, truck, design fee and prototype testing, the small abattoir cost a total of $150,000. While this sounds high, Dunlop said it was less than just the holding corrals and the waste handling system would have cost in a fixed site plant.

One factor that increased the cost was that the plant was designed to be completely self-sufficient. It generates its own electricity, carries and heats its own water and chills and stores on-board the carcasses of the animals it processes.

Dunlop said that a unit that uses the farmers' power and a separate truck for hauling the carcasses could be built for approximately half of the San Juan's prototype.

A non-ferry unit could also be taller and longer. A unit he designed for a California graziers' co-op is two feet longer and six inches taller. Height is very important because USDA regulations prohibit the carcass touching the abattoir floor.

---

**Price Makes a Big Difference**
**Most grassfed beef sells for between $2.50 and $3.50/lb to the consumer.**
**A 660 lb carcass at $3.50 is $2310**
**A 660 lb carcass at $3.00 is $1950**
**A 660 lb carcass at $2.50 is $1320**
    **Production cost is the same for all three carcasses, but the highest price shown brings nearly $1000 a head more.**

The abattoir is designed to only skin and gut the animal. The carcass is then taken to a mainland site for final fabrication. Again, due to ferry height restrictions the carcasses must be quartered to prevent their touching the floor in the chill locker.

On most farms and ranches, a temporary corral is built in a clean, grassy pasture. USDA regulations require that the animals be held at least 50 feet from the abattoir.

Animals to be harvested and animals that will not be harvested are penned together. *The non-harvest animals are there to keep the last animal to be harvested calm.* Dunlop said that being alone is highly stressful to herd animals and is particularly terrifying to sheep.

The animals to be harvested are first inspected by the USDA inspector and judged as to whether or not they are healthy. The animals are then stunned and bled outside the abattoir one at a time by the butcher. The animals can be stunned either through the use of a spring loaded bolt or a .22 rifle.

The bolt is preferred for stunning as it allows the head to be sold to the consumer.

If a rifle is used, the USDA will not allow the head or tongue to be sold for fear of lead bullet fragments in the meat. *Whole heads are prized festival food in Hispanic communities* and each is worth at least $50. Therefore, saving this one piece of meat almost pays for the animal's entire harvest cost of $75.

Once the animal has been bled, it is pulled up a slide into the abattoir by its rear legs with an electric winch. Another winch is attached to the front legs and the animal is lifted onto an open sided, metal cradle for evisceration and hide removal. Once the hide is removed the door to the abattoir must be shut before the carcass is penetrated with the knife.

Johnny Johnson, the USDA inspector, dissects the animal's lymph nodes, head and liver for signs of disease. He also tests the meat for the presence of E-coli. He told me he has never found any E-coli or had to condemn any livers from the grassfed beeves. He said the consistent good health of the grassfed animals has made him a believer in the program.

The hides are stored and shipped off-island in large plastic garbage cans. The offal is collected in a front-loader and goes onto a farm wagon for on-farm composting and return to the land.

Dunlop said with one butcher, the abattoir can harvest and process five beeves a day or twice that many sheep or hogs. He said the slow kill rate has not been a problem as most co-op members seldom want to harvest over three to five beeves at a time. The daily harvest rate could be doubled with two butchers but much of the butcher's day is spent on the ferry getting to work rather than at work.

"These are high priced people," Dunlop said. "You want to use them as fully as possible."

Currently, the abattoir is used to harvest animals two days a week and the butcher fabricates that same meat off-island the other three days.

The co-op charges $75 to harvest a beeve and 60 cents a pound for fabrication. Lambs are $30 for harvest and 80 cents a pound for fabrication.

The real bottleneck for the co-op is not the abattoir but in the fab plant and in finding cooler space for aging. An ideal situation would be multiple portable abattoirs funneling carcasses to a high volume regional fabrication facility.

He warned that no USDA-inspected abattoir could be as cheap as a custom plant.

"It just costs more to do it the USDA way," he said.

---

**Production Cost Estimate for 1100-lbs Steer**
**$350 Harvest, age, cut and wrap**
**$100 Marketing direct costs**
**$550 Production cost (Birth to Harvest)**
**$1000 Total cost**

However, he felt that the increased costs were more than offset by the increased market a USDA-inspected, portioned product could bring a grazier.

"Once grassfed beef is available by the cut in your community, you'll never be able to sell another half or quarter."

You can't hope your local abattoir hangs around in case you need him. You have to use him to have him."

## After Harvest

Allen Williams works as a branded beef consultant with the Jacob Alliance and has several grassfed beef clients.

He said the USDA approval of portable abattoirs will be a huge benefit to the development of grassfed beef where most operations are small. These rolling, small abattoirs allow the animals to be harvested in the pasture with no stress. The skinned and eviscerated carcasses can then be transported to a large fixed facility for aging and fabrication.

Williams said *the limiting factor in all niche market beef production is cooler space for aging.* Even a relatively small kill floor could kill 50 to 80 beeves a day. The problem is where to hang the carcasses while they age, which is why high volume commodity beef is not aged.

With dry aging the carcass is exposed to the chilled air. With wet aging the product is sealed in airtight packaging and then aged. Williams said quality beef marketers should use the dry aging technique as dry aging also concentrates and intensifies the meat's flavor whereas wet aging doesn't do this.

His current recommendation is a minimum of 14 days but some clients extend this to 21 days. Care has to be taken not to over-age the meat as it will turn mushy. Mushiness is a bigger problem with wet-aging than dry aging

He said that the "mouth feel" of the meat was very important in a quality eating experience. Consumers don't like a too soft bite anymore than they like a too tough bite.

He said that while aging and mechanical tenderization can help with a tough carcass they were still just crutches. A far better solution was to breed genetically tender cattle. "Our whole beef technology is oriented toward trying to improve crap," he said.

Williams said many commodity cuts are now injected with a saline and rosemary solution to try to improve tenderness and flavor. This means that in every 10 lbs of commodity beef the consumer buys there is 1.2 lbs of water.

Rosemary is used to try to give relatively tasteless grainfed beef some flavor.

"Crutches like water injection, extended aging and mechanical tenderization would be totally unnecessary if we took

---

**Blade Tenderization**

A mechanical tenderizer has spring loaded tiny knives that penetrate the meat but not the bone. This cuts any gristle or connective tissue that might make the beef tough. The meat is sent through lying flat the first time and then standing on end the second time. This creates a crosscut inside the meat that insures meat tenderness.

The Meat Animal Research Center in Clay Center, Nebraska, said that even with cattle that have been genetically selected for tenderness and have well marbled meat, all briskets and flank steaks will still probably be too tough for most consumers and should be mechanically tenderized routinely.

With generally tender cattle, the top sirloin, shoulder clod, chuck roll, top round, bottom round and the sirloin tip should also be mechanically tenderized to insure customer satisfaction.

It requires approximately five minutes to mechanically tenderize a hundred pounds of meat. Most stores and abattoirs won't mechanically tenderize meat because of the extra time it requires even though it insures 100% customer satisfaction.

better care with the cattle genetics we use."

He said grassfed round steaks from genetically tender cattle were as tender as commodity ribeyes and could be grilled like a ribeye.

Currently one out of every four American beef cattle are too tough for a branded beef program. He said a big part of the Jacob Alliance's job is weeding out these tough cattle before they make it to the kill floor. The way they do that is with Ultra-sound imaging.

He said that today's Ultra-sound technology was 99% effective in discovering tough eating cattle (high connective tissue). These tough cattle can be identified and culled at weaning with the Ultra-sound before a lot of time and grass are put into them. Even better, the Ultra-sound can be used to weed out seedstock and brood stock with tough-eating connective tissue.

The Ultra-sound software the Jacobs Alliance uses detects meat tenderness, backfat thickness, quality grade and ribeye shape and size. Having a consistent size and weight rib-eye is important for marketers selling through E-commerce as consumers want steaks they buy to all be about the same size.

---

**Electrical Stimulation**
**Electrically stimulating the carcass moments after slaughter has caused muscle pH to plummet.**
**Other attributes of electrical stimulation are:**
- **A 23% increase in tenderness.**
- **Enhanced flavor.**
- **Brighter red muscle color.**
- **More pronounced marbling. (Easier for the grader to see.)**
- **An eight percent increase in beef quality.**
**"I believe that electrical stimulation can have a big effect on the eating quality of grassfed animals," said Jo Robinson.**

He said it is absolutely essential that backfat thickness be at least 25/100ths of an inch to maintain tenderness while cooling the carcass.

In addition to tenderness, genetics can make a big difference in the retail yield percentage. Depending upon the muscle/bone ratio, there can be as much as 100 pounds difference in retail meat yield between two animals weighing the same liveweight.

"At (grassfed client) Burt's Beef's wholesale price of $3.90 a pound, that's a difference of $390 a head."

Williams said that luckily *all carcass traits were highly to moderately heritable.*

However, research at the ARS Roman L. Hruska Meat Research Center (MARC) has found that meat tenderness is only 50% due to the genetics of the animal. The rest is the result of non-genetic factors such as stress and diet.

## How Tenderness Occurs

Research by Mohammad Koohmaraie and feed technologist Tommy L. Wheeler in the 1980s found that the tenderness of meat changes during the port mortem aging — first going through a toughening phase before the tenderization phase begins.

Right after slaughter, meat is tender. But for the next 12 hours or so, rigor mortis takes place, stiffening the muscles and making the meat tough. Toward the end of the toughening phase, while the carcass is hanging in the cooler, the tenderization phase begins, which makes most meat — but unfortunately not all — suitably tender.

The researchers found that steaks sold before 14 days of aging are more likely to be tough. Such long aging is very costly due to the large amount of cooler space this requires and has been resisted by the commodity beef processors.

Even worse, the 14 day aging still does not "guarantee" a tender steak. Some will still be unacceptably tough.

It was hoped that a genetic "magic bullet" could be

found that eliminated the need for aging.

MARC discovered that tenderization was caused by the enzyme u-calpain, which degrades some muscle proteins. However, u-calpain requires calcium to be present to do this.

Koohmarie and Wheeler have developed a process of calcium marination that has improved meat tenderness and juiciness.

Again, this has been resisted by beef processors as it slows down their inventory turnover, but might fit at small abattoirs.

Subsequent research discovered that while u-calpain causes the protein degradation that improves meat tenderness, it's actually the activity of a protein called calpastatin that determines how much calpain is active — and thus how tender the steak will be.

However, attempts to develop a tenderness classification system based upon calpastatin activity have not been successful. This is because calpastatin explains only 45% of the variation in tenderness, which is not high enough for accurate classification.

---

**Tenderness Factors**

- **Genetics — reduced level of calpain enzyme.**
- **Age of animal — Less than 36 months.**
- **Stress — Muscle pH (7.5 or lower).**
- **Electric shock — Little effect on tropical breeds.**
- **Aging — Six weeks best, but not all carcasses respond to aging.**
- **Marinades — Papaya best, Kiwi next, but will only tenderize the meat surface.**
- **Calcium — High blood calcium levels improve effect of aging.**
- **Hydrodyne — Explosion in water, the ultimate tenderizer.**

**MARC, Clay Center, Nebraska**

---

A complementary genetic program has tried to identify gene markers for u-calpain. This has been thwarted by the heavy use of crossbred beef animals in the USA, but u-calpain gene markers have been developed for the most popular purebred beef breeds.

However, research indicates that there may be many genes other than u-calpain and calpastatin that influence tenderness — albeit at lesser effect. As a result many more genetic markers will be required to explain enough of the variation to accurately guide breeding choices. "There is only so much we can do with genetics, " chemist Timothy Smith of MARC said. "The rest is determined by how the animal and the meat are handled throughout the various steps of beef production."

It also has a lot to do with how well the animal is fed.

Throughout history, humans have sought out fat animals in preference to lean ones. *A major problem with much of the grassfed meat being sold in America today is that it is not properly fattened (finished).*

The tiny fleck of fats in a marbled meat insure that the consumer gets some fat with each mouthful. This has several major benefits:

■ The marbling fat contributes the majority of the beef flavor. Protein is tasteless. Only fat has flavor.

■ The marbling fat ensures that the meat stays moist during cooking.

■ The marbling fat stimulates the salivary gland to keep producing saliva during chewing and so help promote the moist juiciness sensation.

■ The marbling fat provides lubrication for the teeth biting through the meat. As a result, well marbled meat "feels" tenderer in the mouth even when it may not be.

■ Marbling fat produces the lactic acid necessary to rapidly drop meat pH and prevent spoilage.

■ The healthful CLA that is attracting many consumers is located in the fat and not the protein portion of the meat.

Research at the University of Nebraska found that when

two pieces of meat with equal tenderness (as mechanically tested) are used in preference trials, the meat with the most marbling will invariably be chosen as the tenderest. Apparently, the human mind connects juiciness with meat tenderness even though there is no physical correlation

Dr. Tommy Wheeler of the Meat Animal Research Center in Clay Center, Nebraska, said that while marbling does produce superior meat flavor and provides a natural lubricant that makes the meat easy to swallow, these positive attributes are achieved by meat in the USDA Select grade and that fattening to Choice was not necessary for eating quality.

"With Select grade cattle tenderness is the only major economic variable in consumer satisfaction," Wheeler said. "Flavor and juiciness are considered to be adequate."

With so many variables affecting beef tenderness, Wheeler said MARC had come to the conclusion that *the only way to guarantee a tender product was to test each individual carcass.* At MARC this is done with high tech equipment to speed throughput and to limit the human variability factor, but this high priced technology is not absolutely necessary.

"With the small numbers of animals the typical direct marketer is killing at a time, meat tenderness testing can definitely be successful using a low-tech approach," he said.

The ribeye has been found to be the best predictor of overall carcass tenderness. Tests using lower cost cuts were found to not be reliable predictors of the tenderness of the higher grade cuts.

At MARC the carcass is cooled and then cut at the 12th vertebra (the same place the carcass is split for grading), and a one inch ribeye steak is cut out. This steak is then precision cooked to medium. A narrow slice of meat cut longitudinally with the grain of the meat is tested in a machine that measures the resistance of the meat to being mechanically torn apart.

Wheeler said the estimated $200,000 cost of this apparatus is steep for the low volume abattoirs most direct marketers use. Therefore, cooking a ribeye to medium and eating a slice of it could substitute. "Most of us know a tough or a tender steak when we chew it. If it is tender enough for you, it will probably be tender enough for your customers."

This testing can be done at any time after slaughter. The longer the time before the test is done after slaughter the more the aging will affect the meat's tenderness. He said that carcasses that are found to be tough can be salvaged through mechanical tenderization and sold at a lower price to non-gourmet markets.

Unfortunately for direct marketers who are usually selling the whole carcass, even carcasses where the higher value cuts test tender will have some non-tender cuts. He said for consumer satisfaction the brisket, flank, cube steaks and round cuts should always be mechanically tenderized. This can be done with a blade tenderizer or a Jacard tenderizer.

**Generally Tender Cuts\***
All Tenderloins
All Top Blades
Most Ribeyes
Most Strip steaks

**Mostly Tender Cuts**
Top Sirloin
Shoulder clod
Chuck roll
Top round
Bottom round
Eye of round
Sirloin tip

**Probably Always Tough**
All Briskets
All Flank steaks

\* From a carcass tested and identified as "tender."

## Other Tenderness Factors

It appears we are ever so slowly unraveling the Gordian knot of meat tenderness factors. Most, but not all, of these are related to how fat the animal is at harvest.

In all other tenderness issues it was marbling fat that has been the key fat component, however Jo Robinson has found research showing that outside carcass fat was also a key component. However, this was not due to any inherent eating quality issue but due to the protection it offers from man-caused technology problems.

Robinson, as the editor of www.eatwild.com spends most of her working day combing the Internet internationally for vetted scientific research on grassfed meat and dairy products that she can report on her website. This includes back issues of scientific journals that are now posted on the Internet.

While she mainly concentrates on human health issues, she is also intrigued by post-harvest techniques that increase meat tenderness. She recently found some late 1970s research from the University of Wisconsin that appears to be hugely significant to many grassfed meat producers and I asked her to present it at one of our schools in Sacramento.

One of the persistent questions I have continually asked myself is, "Why is North American grassfed beef so much worse than the rest of the world's?"

In virtually no other country is meat tenderness the issue it is in the United States.

I get e-mails from overseas readers questioning this obsession all the time.

"We don't have these tenderness problems," they say.

And that appears to be largely true.

I have had steaks from grassfed Zebus in Brazil, Paraguay and Central America that were as tender as American grainfed Angus. And, none of this meat had any marbling fat.

You can cut most of New Zealand's table grade grassfed beef with a butter knife. In fact, special serrated-edge "steak knives" are seldom provided at their better restaurants.

As I have told you before, the absolutely best steak I have ever tasted came from a nine-year-old French Aubrac cow and the best meat I have ever eaten was from a 20-year-old bison cow in Hawaii. Both of these examples were very tender thanks to slow cooking and had incredible flavor.

Even here in North America, wild venison and elk is seldom as tough as many of our grassfed steaks. Why? Why? Why?

The following may be an answer.

Again, as in most eating quality issues, it all comes back to fat. In this case, backfat.

Apparently, according to the Wisconsin research Jo found, *the biggest factor in meat tenderness has to do with how fast you chill the carcass in the first two hours after harvest.* This research was done by J.V. Locner, R.G. Kaufman and B.B. Marsh at the Muscle Biology and Meat Science Laboratory at the University of Wisconsin in the late 1970s.

The research was published in "Meat Science 4" in 1980 and is currently posted on eatwild.com if you would like to review the full text.

In the 1970s, due to the high grain prices at the time, the federal government funded a lot of research on grassfed beef production. Apparently, this Wisconsin research came from that funding.

Here's the gist of the problem:

## Slow Chill Crucial with Lean Meat

In commercial plants, the chiller is set to cool the fattest carcasses likely to be encountered.

The Wisconsin researchers found that the fatter and the larger the animal the slower the carcasses chill in the freezer.

Smaller, leaner grassfed carcasses chill much faster. Particularly, if they are cut into quarters as many grassfed animals currently are.

This also applies to small animals such as goats and sheep.

In the Wisconsin research, Angus calves sired by the same bull were put on two different feed regimens.

Half were fed corn and corn silage and the other half alfalfa hay for between 170 and 190 days.

The corn-fed cattle weighed around 1250 lbs at harvest and the hay-fed 950 lbs.

The corn-fed cattle gained an average of 2.2 lbs and the alfalfa-fed cattle gained only a half a pound a day.

Yield grade for the hay-fed cattle was USDA 2 but a whopping 4.7 for the corn-fed cattle.

After harvest, the carcasses were halved with the right sides of each group going to a chiller set to - 2°C/29°F and with forced air movement and the left sides to a chiller set at 9°C/48° F with no air movement.

A thermocouple was placed in the longissimus muscle opposite the 12th thoracic vertebra to continuously monitor carcass temperature changes.

After 7.5 hours post mortem, the carcasses were all transferred to the -2°C/29°F cooler to prevent excessive microbial growth.

At 24 hours post mortem, the thermocouples were removed and all sides were moved to a 3°C/36°F cooler where they were ribbed and evaluated for USDA quality and yield grades.

A 25cm section of the cranial end of the short loin was cut from each side, wrapped and left at 3°C /36°F for an additional 48 hours before being frozen and stored at -20°C/ -5°F until taste and shear tested and estimated for marbling fat.

A sensory panel found both sides of the hay-fed beef inferior in tenderness, juiciness, flavor and overall desirability.

The fact that both carcasses were found to be equally tough ruled out any influence in cold shortening. Something else was at work here.

Here's an example of cold shortening.

If you've ever jumped into an ice cold swimming pool on a hot day, you have felt your muscles reflexively contract.

If muscles go into rigor mortis in such a contracted state, you get a tougher piece of meat than if they were relaxed.

Most researcher have attributed some of the tenderness differences between lean and fat animals to cold-shortening. However, the Wisconsin research found cold-shortening was seldom a problem except with very small, ultra-lean carcasses.

The Wisconsin researchers found the meat toughening problem occurs long before rigor mortis sets in. In fact, it appears that it is ***the muscle temperature in the first two hours after death that is the most critical*** for meat tenderness.

At two hours post-harvest, the internal muscle temperature of the grainfed animals was still around 38°C /101°F whereas both of the lean carcasses were below 34°C/94°F with the rapidly chilled lean side down to 28°C/ 82°F.

In comparing the two fat carcass halves it was found that there was only 1°C/4°F in temperature difference in the meat muscle despite the huge temperature differences in the chiller. The grain fed carcass' fat cover effectively slowed the cooling of the meat muscles.

Now, here's what's really interesting.

The taste panel could detect significant tenderness differences with the 1°C /4°F higher temperature half being found more tender. Remember, the hay-fed carcass' muscle was 10°C/50°F cooler!

## A Hot Box for Tender Beef

Previously published research in 1950 cited in the Wisconsin study had shown that the best way to create tender meat was to put the carcass in a "hot box" rather than a chiller and to hold the internal muscle temperature at 98.6°F/37°C for four to five hours.

The Wisconsin researchers found that the tenderness temperature sensitivity actually occurred in the first two hours post-harvest. How rapidly the carcass cooled after the initial two hours was found to be insignificant to the subsequent tenderness of the meat.

"The tenderness superiority of well finished over poorly finished beef is due largely to differences in cooling rate, but only rarely is it due to differences in cold shortening," the researchers wrote.

So, if it's not cold-shortening what's going on here?

As noted by the MARC research, natural enzymes create meat tenderness immediately after harvest and calcium is the catalyst mineral for these enzymes

Wisconsin researchers speculated that these natural tenderizing enzymes may also be stimulated by heat and possibly retarded by cold. They noted that such rapid internal chilling does not normally occur in nature.

Their conclusion was:

"It is now clear that the superior eating quality of the fatter animal is caused not by the prevention of cold shortening, but by an as-yet unidentified mechanism that is promoted by the maintenance of near physiological temperatures during the first 2 to 4 hours after slaughter.

"Logical extension of this conclusion leads quite simply to the prediction that the tenderness of lean beef could be enhanced appreciably by application of a high-temperature, short-time heat treatment, thereby stimulating the 'natural' slow initial cooling of larger and fatter sides."

Now, this research seems to give some answers to my previous questions.

The researchers noted that North American chillers are set to chill the "fattest" carcass the abattoir is likely to encounter. In other words, North American chillers are set for large, fat carcasses that are very difficult to chill.

This is devastating to the tenderness of leaner carcasses

---

**Customers are Willing to Pay**
**25% premium for a tender steak.**
**30% premium for a "guaranteed" tender steak.**
                                        **Virginia Tech**

in the same chill room.

I will wager that the chilling rate in Brazil, France and New Zealand is considerably slower than ours because their maximum external fat level is much lower.

When I was in Argentina several years ago, I went through a large EU-inspected, French-designed "export" abattoir. The carcasses were all shrouded in salt water-dampened muslin shrouds immediately after slaughter. They said the salt water was to prevent microbial growth on the carcass and the shroud was to slow carcass cooling.

The carcass chain (visualize a mono-rail track with whole disemboweled carcasses hanging from a two-wheel truck) went in a huge nearly 360-degree loop before entering the chiller.

This chain was extremely slow moving. So slow in fact that its movement was almost imperceptible to me.

I would guess that the animals took at least an hour to reach the chiller and maybe longer. As I remember it, these shrouds were removed as the carcass entered the chiller.

Do they know something or was this initial slow chilling serendipitous luck?

I suspect since this was a French built unit, it was the former. However, I suspect in much of the world it is the latter.

Refrigeration is expensive and cooling faster than absolutely necessary would be avoided.

In much of the world, the carcass is not chilled at all but is eaten "fresh."

I saw outdoor "cooling" of beef carcasses at abattoirs across Central America.

The reason game is more tender may be because it seldom gets to a chiller in less than two hours.

In fact, the tradition in much of rural America is to let the carcass "cool" overnight with its hide on before skinning and portioning the next morning.

This delay in cooling may be why field harvested bison are more tender as well. They usually are transported with their hides on until they reach the abattoir which may be a considerable distance away.

So, what can we legally do about this problem today?

Allen Williams told me that if your animals are properly finished and have at least 3/10ths of an inch of backfat this quick chilling problem will not be a problem for you.

He said if you are using a large volume USDA inspected abattoir getting animals properly fat is probably your only option as they are not going to make modifications for a small volume customer.

However, the reality is that most of our grassfed beef is not properly finished but is on the market. Therefore, I think we should try to do everything possible to eliminate this quick chill problem.

Allen said that if you are using a low-volume custom abattoir you might have some slow chilling options.

Some proposed solutions for lean carcasses by Jo Robinson are carcass shrouds and insulating bags, having your animals harvested late in the day after the chiller has been warmed by earlier carcasses, leaving the carcass whole rather than halved or quartered for the first two hours and making sure the carcass does not have a chiller fan blowing directly on it.

In a non-USDA inspected, low volume abattoir you might be able to talk them into delaying putting the carcasses in the cooler for two hours.

Unfortunately, most USDA inspectors are trained in high throughput plants and believe that chain speed is important in preventing microbial buildup.

Of course, it appears that the two-hour "hot box" treatment would be the ultimate solution. However, this would have to be proven to the USDA as "safe" and will require vetted university research to prove this does not create a microbial hazard to the consumer.

Jo found a 1982 research paper by K.E. Wilhelm and B.B. Marsh titled "High-temperature tenderization of beef sides: bacterial considerations."

In this research, lean beef sides were held at 37°C (98°F) for three hours before being transferred to a chiller. The researchers found "no significant increase in bacterial numbers or a proliferation of potential pathogens or food-spoiling organisms."

The USDA abattoir guidelines (HAACP) specify that the surface temperature of the meat must be below 40°F within 24 hours. Jo said this should not be a problem with the faster chilling grassfed carcasses.

She said Washington State University is interested in doing research on this and is looking for grant funding.

**Flavor**

After tenderness and marbling, the next big issue is flavor. Most North American consumers are used to the extremely bland, almost flavorless taste of grain-finished beef. While this is not something we should aspire to, we need to be careful with objectionable off-flavors.

A volunteer taste panel of 80 people ranging in age from 18 to 60 at Auburn University turned *a definite "thumbs down" to ribeye steaks from steers that had been finished on fungus-infected fescue*, but a "thumbs up" to steers finished on annual ryegrass and fungus-free fescue.

The fungus infected fescue meat was compared to meat from steers finished solely on annual ryegrass, annual ryegrass and hay, fungus-free fescue and a grain based feedlot diet. The steers were all near 1050 lbs in weight and were Angus and Angus X Hereford crosses.

The taste panel rated the steers finished on annual ryegrass as the most similar to those finished on grain with the overall flavor variation being considered statistically insignificant.

The ryegrass and hay steers were ranked as similar to the ryegrass-only steers but were rated as slightly less tender.

The steers finished on fungus-free fescue were rated slightly less favorable than the feedlot-finished steers in taste, tenderness and overall preference but similar to the ryegrass-finished beef.

The beef finished on fungus-infected fescue were rated inferior to feedlot-finished beef in all traits except aroma.

The consumers slightly preferred the aroma of ryegrass and fungus-free-fescue beef to all other treatments including feedlot.

The various finishing diets did not influence moisture, fat,

---

### Real Beef Flavor Versus Veal

**Animals that are killed lighter than 65 to 70% of their mature bodyweight lack the ability to marble because they have not yet developed those fat cells. This immature meat can be tender but it will be tasteless since "beefy" flavor is actually the taste of the fat. Australians quite correctly call this class of young unfinished animals "veal" because that is what it tastes like and how it should be served. Veal requires sauces, gravies and/or seasoning to obtain any real flavor.**

**Most Americans tend to think of veal as 150-lb male dairy calves. A beef animal with a mature weight of 1500 lbs cannot develop any marbling fat cells until it weights nearly 1000 lbs. In other words, you can have 1000-lb vealers if you are using large phenotype breeds. Some people like the tastelessness of veal and will pay a premium for it. That's good because with vealers you are going to need a sizable price premium compared with finished beeves. Here's why.**

**Young unfinished beeves typically have a yield of only 50% versus 60 to 65% for physiologically mature and properly "finished" beeves.**

**If you are selling your beef direct to the consumer for $3.00 a pound, a 10 to 15% difference in carcass yield is several hundred dollars per head difference in your income.**

or ash concentration of the ribeye muscle.

Protein concentration of steaks from cattle finished in the feedlot was slightly higher at 23.1%, compared to 22.1% for ryegrass cattle and 22.4% for ryegrass/hay. However, this small difference was said to be of little nutritional significance.

The color of the meat was similar for all treatments but the feedlot-finished cattle had slightly lighter colored fat.

Analysis of samples for B-carotene and vitamin A indicated significantly higher amounts in the grass-finished animals.

The B-carotene in ryegrass steak was 44% higher than the feedlot beef and 63% higher in the ground beef.

Vitamin A as measured in the liver was 81% higher in the ryegrass beef than in the feedlot beef.

The research was done by Dr. David Bransby and Amy Simonne.

The researchers concluded that grass-finished steers can be produced well within the range of mainstream consumer acceptability with the possible exception of those from fungus-infected fescue.

## Wheat Pasture Plus Brassicas Can Taint Meat

Wheat pasture and brassicas can taint the flavor of both meat and milk in grazing animals according to New Zealand research. Brassicas are forages such as turnips, rape and kale.

The taint problem depends upon the number of days fed before slaughter and the feed. For example, brassicas are worse meat flavor tainters than wheat.

To ensure good flavor, the animals should be removed from wheat or brassicas at least three weeks prior to slaughter and moved to ryegrass.

Other products found to impart an unpleasant taste are seedweed derived products such as kelp, flax oils, and certain weeds.

Jo Robinson told me that what really hit her most in the West Coast chef tastings she supervised was that in every instance the winner thought his beef was not as good as it should be while the losers all thought their beef was "good enough."

And, I predict that these are the very attitudes that will eventually separate the winners from the losers financially as well.

If you want a long profitable run in this business, I urge you to continually strive for gourmet and not settle for good enough.

# CHAPTER 14

# Pioneering a New Industry

A couple of years ago, my wife Carolyn and I were in Honolulu for a convention. While we were there we visited a waterfront museum that had a replica of the frail raft that the first human settlers to Hawaii used for their long 1800 mile journey across the open ocean.

"What kind of person would take such a huge risk?" I asked the museum curator.

"A king without a kingdom," he replied.

I thought his answer perfectly encapsulated the pioneer spirit. *If you have nothing, you can still be king if you get there before anyone else.*

The whole history of human economic progress is dependent upon a tiny minority of irrational people totally driven by their dream that forces them to "go where no man has gone before."

Harvard professor, Michael Porter, in his finance text-book *Competitive Strategy* said that all new industries have certain common characteristics. I have taken Porter's list and interpolated them into a grassfed context to illustrate the anxiety you are most likely feeling about joining a pioneer industry. You'll see that this is entirely normal.

Porter said the common characteristics of pioneer industries were the following:

## 1. Technological Uncertainty

In a new industry no one is quite sure the best way to produce the new product.

We can see that in today's grassfed industry — some people are trying to produce fattened beef on cow-calf quality pastures. Some have even made non-improvement of pastures (native range) as a part of their marketing statement. Others are convinced they must have some sort of energy supplement and cannot do it on grass alone.

This is the technological uncertainty Porter was talking about.

People start out trying to do something new with the skills they presently have. However, when we change to producing an end product we change everything including the skill set needed. This is very unsettling to people whose persona is based upon their current competence.

## 2. Strategic Uncertainty

In an early stage industry, there is no one "right" marketing strategy. Some people swear by farmers' markets while others swear at them.

---

**Twenty years ago I thought the idea of "farming" grass with the same care as a corn crop and harvesting it with a four-legged "combine" steered with minimalist electric fences would be an idea that would sweep America in two weeks.**

**New Zealander, Vaughan Jones, said he thought it would take two years.**

**Neither of us ever thought it would take 20 years — a whole generation — to finally arrive at the beginning of the beginning as it now has.**

**Every day I am reminded of Joel Salatin's observation that no pioneers ever did it for themselves. They all did it for their children.**

Compounding marketing difficulties is the fact that the product's consumer may be philosophically and economically very different from the pioneer producer. Market and customer research is usually very limited in the startup phase. This is particularly the case as to who one's true competitors are.

For example, *the major competitor to grassfed beef in most alternative food markets is not a grainfed meat but a pseudo-meat product made from soybeans.*

Porter said the degree of marketing freedom is greatest in the early stage of market development. Some people find this exhilarating. Others find this freedom to make it up as you go along terrifying.

New industries require very short and inexpensive distribution chains. To maintain the wider margin needed for a startup industry, there is little money to feed a full-time distributor. This is why most pioneers have to direct market.

However, eventually developing a distribution chain is essential for the industry to grow. People cannot buy what they cannot find.

### 3. High Initial Costs Due to Low Volume

Porter said that market pioneers had to look at the "potential" cost structure rather than the "actual" initial cost structure. Many of the high costs faced by new industries are the result of low volume. A good example of this is the high abattoir costs currently faced by pioneer producers.

Porter said all new industries face the dilemma of surviving the high costs of the low volume start-up period. However, such volume-caused costs tend to fall rapidly as volume increases.

### 4. Poor Business Skills of Proprietor

Porter said that all new industries are created by people who can create an exceptional new product. As a result, most new industries are started by production-oriented people.

However, once the production model has been perfected

the producer must quickly add the skills of marketing and finance for it to grow.

These skills can be added by hiring consultants, taking a crash program for personal education, by hiring employees with the new skills or by taking in partners who have the needed skills.

### 5. First Time Buyers

Porter said that first time buyers are hard to find and expensive to educate. No one is starving to death in North America so all food product marketing must induce substitution. This substitution effect is created primarily through product sampling and is why new products have to be considerably better than what currently exists.

### 6. Short Time Horizons

The pressure to create a positive cash flow in a new industry is so great that production shortcuts are often taken that hurt the industry in the long-run.

A good example of this is the current widespread practice of harvesting animals that are not fully finished to meet short-term demand. This results in a sale today but the loss of a future customer forever due to the poor eating quality of unfinished beef.

Often newcomers to an industry imitate the practices of the earliest pioneers even when those practices have been proven to have long-term negative implications. A good example of this is attempting to finish cattle in the late summer and fall on perennial forages.

### 7. Seeking Subsidies

New industries are short on cash and so often seek out government subsidies for product and marketing research. Porter said such grants add a great deal of instability to an industry because the recipients become dependent upon the outside funding.

This funding is politically motivated and can quickly disappear under the pressure of a threatened established industry. These subsidies also involve the government in an industry, which Porter said can be a mixed blessing.

## Steep Learning Curve

We are a new industry of gourmet grass-finished beef producers, and we are on a steep learning curve. For most of us, this creates feelings that are a combination of exhilaration and fear. What we need to constantly remind ourselves is that those feelings are normal and have been felt by all pioneers.

Comfortable people don't take big risks. However, sometimes comfortable people are lured into pioneering with the mistaken idea that the pioneer's life is easier than what they currently have.

*The Wall Street Journal* had a special section on entrepreneurship. It pointed out that the widely bandied figure of 95% of all people who start new businesses go broke by the fifth year is completely wrong.

Only a small number go broke. The rest quit.

They quit because they find that they have a far worse job than what they left.

Starting a new business is a U-shaped process. Most people start with a wildly optimistic projection of how easy it will be and then fall to the depths of despair between year two and three. As the *Wall Street Journal* pointed out, this bottom of the trough is where we lose most people whose goal was to create a job for themselves rather than a kingdom.

*Financial breakeven normally isn't reached until year five to six and financial breakthrough doesn't happen until year twelve.*

---

**Most people are willing to spend anything to get what they want except the time it takes to learn how to do it exceptionally well.**

---

The life of ease that initially lures most people into starting a business normally doesn't occur until year 25 or 30.

And, then it comes only if you have learned how to hire and train competent employees.

I have been down this long road myself and am now nearing 30 years of rowing my own boat. And yes, life is good. I owe no money and am able to pay all my bills by return mail without worrying about it. This is what my Dad called the country boy definition of wealth.

My heart bleeds for all of you walking through that scary second and third year when all of your rational thoughts are screaming at you to quit, because I have been there. Perhaps the biggest help I can provide is to frequently tell you that *the pain you are feeling is normal and that it will eventually pass if you persevere and keep trying new things.*

To put this in Nature's context, every new human being comes into this world via a lot of pain on the part of its mother. Whether that pain was really worth it cannot be objectively known by bystanders for 25, 30 or 50 years. And yet your mother knew you would turn out great the day you were born.

A mother can look at that newborn pink blob covered with blood and amniotic fluid and see the fully grown man and woman it will become. The entrepreneur has that same mother's eye about his business baby. He can see the huge success it will grow up to be as clearly as everyone else can see the helpless mess it is now.

I am always concerned when I hear startup operations define their goals in the terms of a level of income. This income level is invariably what they had in their present or former job.

The reason this concerns me is because if you are only in it for the money, you probably won't stick with it through the

---

**Knowledge is worth the most when only a few have it, and this only happens in a new industry.**

tough years when the money isn't there.

What I like to hear is someone whose goal is to completely change the world. *You will never be more successful than your biggest dream.* So big dreams can sustain you and keep you pushing on when small dreams can't.

Because of the glamour inherent in an "artisanal" food product, grassfed beef is attracting a sizable number of what I call "play" entrepreneurs.

A lot of these are bored wealthy people who want to be part of something new and fashionable. Others are well-off ranchers who are intrigued by the high returns per head but only so long as it doesn't challenge any of their paradigms such as their current breed of cattle, calving season, favorite grass or grazing management skill.

While these are certainly interesting and entertaining people, I predict few of them will hang around for the long slog that building a new industry infrastructure requires.

The problem with pioneering a new industry is that the ancillary support businesses necessary to easily get the product from the pasture to the consumer doesn't exist. This support infrastructure is called a "value chain."

### Building the Value Chain

A value chain is all the other businesses that are required to get your product from you to your consumer.

For example, with this book you are holding the value chain that includes the farmer who grew the tree and sold it to the paper mill who sold the paper to the printer who printed the book you are holding to the Post Office who got it to your house. This book could not exist if any one of those other businesses didn't exist.

The lack of this infrastructure is a huge hurdle to people who want to get big in a hurry in a pioneer industry. They arrive on a deserted island ready to be king but are shocked that there is no HBO or air conditioning available. Too often their impatience leads them to try and buy an instant infrastructure rather than let it develop organically on its own.

Again, using nature as our guide, *you can't graft a full grown tree trunk onto a sapling* and speed up its development. The sapling does not have the root system to support such a heavy weight and will quickly fall over and die. *Everything in nature starts small and grows from initial small successes to ever-greater successes.* This is often called a "learning curve."

In business, growth that occurs from retained profits is called organic growth in recognition of its similarity to nature's growth pattern. However, this is too slow for people with gobs of capital. These people are willing to spend anything but the time it takes to build a sustainable business.

*In every new industry there is a correct sequence of infrastructure development that must be followed.* Most people planning a branded beef program mistakenly identify the lack of a kill and fabrication plant as the primary choke point.

The failure of several recent new large-scale, kill-plants built for branded beef programs across the country have shown this is not the place to start.

No, the primary choke-point in all branded beef is the paucity of producers willing to take a risk on producing something different. As I've just discussed, true pioneers are always in short supply.

This is why contract production, which minimizes risk to the producer and can enforce a production model template, is always the fastest way to ramp up scale of production.

Is it still possible for a couple to bootstrap themselves into a ranching career today? Jon and Wendy Taggart, who

---

**Who Makes Money, How and When?**
- **The Introductory Phase — the Producer**
- **The Growth Phase — the Marketer**
- **The Mature Phase — The Financier**
- **The Decline Phase — The Bankruptcy Attorney**
  **Thomas A. Stewart, Harvard Business School**

were profiled earlier, said the answer is yes *if you stay away from land ownership.*

Jon said he was a city boy from Fort Worth who got the ranching bug early in life from weekend visits to ranch children friends. He started out with a cow-calf operation and a short-term ranch lease.

Jon said they currently net between $600 to $1000 per head after all costs. They figure their current home ranch can produce 500 grass-finished beeves a year at maximum capacity.

## Pioneer Marketing

Michael Porter said pioneering markets offer significant opportunities to smaller producers like the Taggarts.

"In an emerging industry there are no rules to the game. This is both a risk and a source of opportunity," he said.

"No rules and no advantage to scale allows very small players to participate in the market. Large players will not participate until the market size has been clearly demonstrated and proven."

*The beauty of being a market pioneer is that you get to be the first harvester.* This means you get to pick the high-margin, easy-to-pick, low-hanging fruit.

I define the low-hanging fruit in meats as any market that will accept a frozen meat product.

Some of these markets are:
- Farmers' markets
- Alternative food co-ops
- On-farm store customers
- Direct delivery customers
- Internet and mail order customers

All of these have a simple distribution chain that retains almost all of the consumer's dollar by the primary producer. These are definitely the best markets with which to start a marketing program.

The beauty of a frozen product is that you can limit your harvest to late spring and early summer and yet sell meat year around. This allows the use of perennial forages, which keeps agronomic costs low and requires only moderate grazing skill. The problem, of course, is that if you sell out you are out until the next year.

From what I have personally seen, most farmers' markets still don't have a grassfed meat purveyor. Graziers on the West Coast get frequent calls off of the eatwild website from New York City, so there is still a lot of easy-to-pick fruit out there. ***But it is not limitless.***

I have written a book on marketing meat through farmers' markets called *Farm Fresh, Direct Marketing Meats & Milk*. However, here is a farmers' market marketing story that isn't in it.

### Selling Meat Like Candy

When you are direct marketing grassfed meat for between $6.50 and $10 a pound how do you overcome the consumer's initial "sticker shock?"

Arkansas grazier, Ed Martsolf, who sells at the Little Rock, Arkansas, farmers' market said the answer is simple: Stop selling it by the pound and start selling it by the ounce.

"Forty cents an ounce is not nearly as intimidating as $6.40 a pound," he said. Martsolf said most people can't do the math in their head, but to head them off he simply holds up a small Snickers candy bar.

"It's roughly the same price per ounce as this candy

---

**To be Successful, Pioneer Products Require:**
- **Alignment with the "right" set of customers.**
- **A small organization that can be excited by small sales.**
- **A plan to fail early and inexpensively.**
- **The discipline to only develop markets that value the existing attributes of the product.**

**Clayton M. Christensen, *The Innovator's Dilemma***

bar," he tells them, and then quickly adds, "However, this grassfed meat is high in Omega-3 and CLA and is a lot better for you than this candy bar."

He said this usually ends any price quibbling.

After 20 years of meat marketing, Martsolf has learned a few things about consumer behavior.

■ One idea is to put your product in a frame of price reference as he does with the candy bar.

■ Another is to have your sale portions small.

"People don't care what your product costs as long as the package price doesn't exceed ten dollars. People come to a farmers' market looking for something for that night's meal. They don't come looking to buy a freezer-full of meat."

■ He wants everything he sells to fit into his hand so he can have the candy bar in the other for the value comparison. This means all cuts are sold boneless.

"Selling boned out meat lowers the yield but just forget about that."

■ He also keeps his meat cuts simple and realizes that most of them are going to be grilled rather than roasted. All meat is sold frozen.

He said to keep in mind that *your primary product is your farm and its story.*

"We are marketing a unique combination of white lab coats and nostalgia. We want to *use the latest medical research but put it into the terms of Grandmother's farm.*"

This marketing of the farm first allows you to expand your market offerings under the same marketing umbrella.

For example, Martsolf sells grassfed beef, lamb and farm-raised honey.

He said an ill-advised foray into wholesale marketing through traditional grocery channels with the ranch's honey production was a complete financial disaster. However, this taught him to respect the simplicity of the farmers' market model as a marketing tool.

The other rapidly growing marketing tool many are using

with grassfed beef is the Internet. Jo Robinson will run a listing of a grassfed offering on her website eatwild.com indefinitely for only $25. This is the ultimate bargain.

Eatwild frequently updates the latest research on the health benefits of grassfed products as well as providing consumers easy access to hundreds of vetted research results. The website has recorded one million "unique" (not repeat) visitors.

The website has a State and Province map whereby consumers can click on their area and see who has grassfed products for sale. They can then click through directly to that producer's website for more details.

She said Internet sales of grassfed products were already significant and growing rapidly. She said *one Missouri ranch currently sells between 60 to 80 grassfed beeves a month on their website* and a great many others sell between six and eight animals a week.

Here is Jo's advice about creating a successful grassfed website.

■      Never, ever use a picture of an animal grazing dead, dry grass on your website.

■      Consumers want to see pictures of green lush grass, fat shiny cattle and thin pretty people.

■      The photos you use are critical because web consumers will typically decide in two to three seconds if they are attracted enough to the site to stay and read the page.

■      A web site home page with too much writing and few pictures quickly turns people off.

■      She said some grassfed home pages are very obscure as to what is being offered and are far too wordy. Put your environmental and personal philosophies on a later page.

■      The primary purpose of your home page should be to let the customer know you are there to sell grassfed products. "Make it very clear how they can do this. A toll-free number is attractive."

■      Pictures of the producer and his family are important to show that this is a real ranch and not a corporate front.

■ If you don't have any children rent some as models. "Pictures of healthy, happy kids on the home page eating your product are good for at least a ten point advantage.

■ Turn everything you are doing into a consumer benefit. An example of this is "USDA inspected for your safety." Robinson said that the safe-ness of the meat is the number one concern of mail-order buyers and must be addressed. A consumer survey found that 70% of Internet consumers would pay a premium for a product with the claim of "Raised on our farm from birth until market."

■ Care must be taken with all photos to send both a picture of authenticity but cleanliness. "Your jeans should be faded in the knees the way work jeans are and not all over like stone washed jeans. All of your clothes should be spotless."

■ Concern about animal well-being is a big consumer concern. She recommends a "humane" label from one of the established animal welfare groups. She said this was particularly important for veal and poultry.

■ A claim of "grazed on 100% Certified Organic pastures" could be clearer than just a Certified Organic label. "Consumers are leaning that many organic meat products are from industrial farms and need reassurance that yours is the real deal."

She said most Internet consumers were typically high income families and wanted to buy from people who appeared to have similar values.

To illustrate this she showed a picture of a family portrait from a grassfed website that showed them sitting on a European-style, dry stone wall.

---

**The most powerful barrier to entry is to do something that makes no sense to the established leaders. Good managers have a hard time doing what does not fit their model for how to make money.**
**Clayton M. Christensen, *The Innovator's Dilemma***

"That stone wall subliminally says, 'We are not rednecks. We are really cool people just like you.'"

While it may take years to fully satisfy all of the existing market for frozen meat, it eventually will happen. Fresh meat is the big market and is something you will have to tackle to pick higher in the tree.

Peter Drucker in his book *Managing in the Next Society* said ***we must always be planning our businesses for the world as it will exist in five years and not as it exists today.*** He said this five year time frame is the minimum time it takes to perfect a new skill or knowledge base.

Fresh grassfed beef on a year around basis can be done but it requires an increase in agronomic and grazing skill. As stressful as pioneering a new industry is, in the long-run it is less stressful than staying in commodity beef.

### The Commodity Trap

The International Monetary Fund calculates that real commodity prices (adjusted for inflation) have declined on average 0.6% a year since 1900. This means that ***every decade commodity producers have six percent less income from the same amount of gross production.***

In other words, if this trend continues, a commodity producer must produce six percent more every decade just to stay even in real income. This has been the source of the "get bigger or get out" syndrome in agriculture.

Unfortunately, the world economy doesn't need six percent more raw materials every decade. The result is an on-

---

**The trick is to plan to fail, but to fail as inexpensively as possible. This is where a small size is beneficial. Your costs are lower so your initial "learning losses" will be lower. Small businesses can also change directions quicker.**
**Clayton M. Christensen, *The Innovator's Dilemma***

going process of triage whereby the most economically marginal producers are continually eliminated.

The big problem in the developed world is that commodity producers are having a difficult time maintaining income parity with the off-farm economy.

In 1960, my father's 200 beef cows produced a taxable income equivalent of $160,000 in today's dollars.

Today, 200 cows are considered a "part-time" income.

The problem is that as the off-farm economy became "smarter" the farming economy has become "dumber." Today, most of the thinking is done off the farm in value-added processing and marketing firms and by farm input suppliers.

My father, an Appalachian transplant to the Mississippi Delta, noted the incongruity that the Delta with a topsoil 60 feet deep produced a far more grinding poverty in its larger economy than the stony mountain soils he had left behind.

Thomas A. Stewart, editor of the "Harvard Business Review" explained this saying, "A wealth of natural resources will be exploited by people with a wealth of knowledge; the value of natural resources is extracted from a place, not created in a place."

Stewart said that *to become wealthy you have to understand what creates wealth.*

And, that is a source of unfair competitive advantage.

He said there are only two sources of competitive advantage. These are: differentiation (no one else has it) and cost (no one else can produce it for less.).

The definition of a commodity is that it is undifferentiated. Therefore, the only way a commodity producer can increase his profits is by cutting costs.

A production innovation that lowers costs can give a temporary competitive advantage to those who adopt it early. However, this advantage only lasts until others adopt it and then not only is the advantage lost but the market price stabilizes at the new lower breakeven price.

In other words, the ultimate beneficiary of all production

innovation is the consumer and not the producer. He said the production innovation cycle in commodities is:

■     Everybody cuts costs.

■     Everybody produces more to cash in on the increased margin.

■     Everybody cuts prices to sell the increased production.

■     Everybody cuts profits.

Stewart said that while cost-cutting can temporarily give you a bigger slice of the pre-existing pie, it can never grow a larger pie. "Those who live by cost-cutting will die by its sword," Stewart warned.

What is considered poverty in the USA would be an above-standard living in most of the underdeveloped world. This means that a developed world farmer is going to lose interest in the game long before one in the underdeveloped world will.

This is the "sword" that Stewart is talking about. Who in North America wants to fight over a day's paycheck that will only buy a bowl of rice?

Similarly, land — that most primary of inputs in agriculture — is now priced out of the reach of commodity farmers throughout the developed world.

Who would have thought 40 years ago that a pretty view, or the number of deer, would trump the agricultural production value of a piece of land by a factor of 10 or more?

Now, remember the hundred-year trend of falling prices. For you to maintain your standard of living selling commodities requires that you increase production by six percent a decade.

---

**Business guru Tom Peters guesstimates that 75% of all new ideas are generated by people new to an industry. Ironically, when you are pioneering, not knowing what you are doing can be a powerful "unfair advantage."**

Can you outbid a knowledge worker in law, medicine or the sciences for that increased acreage? Perhaps, if you maintain an austere standard of living.

The real crunch hits the startup commodity producer. *If you believe that you must own all the land you farm, getting into commodity agriculture from scratch is increasingly hopeless.* The odds that you will sell your farm or ranch to another farmer or rancher are becoming remote except in the most rural areas of the country.

Industries where the young can no longer get in eventually disappear.

An America without farms? Could it happen?

Go to New England and you may see the future that awaits us all. There the land has almost all reverted to forest, and commodity agriculture is almost completely gone.

The same process of reforestation and agricultural abandonment is rapidly taking place in the Southeast as well.

These deflationary pressures are by no means limited to just agriculture. Everyone in the "hardware" business whether they are producing computers, cars or crayons feels them.

"Globalization, outsourcing, and a century of scientific management have made it harder and harder to achieve (competitive advantage) either by means of physical assets, materials and processes," Stewart said.

"There can't be competitive advantage from unskilled work, because anyone can do it; nor can you gain an edge by means of a machine you can buy off the shelf, because anyone can buy it.

*"Competitive advantage comes from something proprietary — or at least hard to duplicate. Uniqueness is where a business wins or loses today."*

The only way to rise above the ever-falling prices of the commonplace is by being different. And, the more different the better.

This is a hard concept for socially conservative rural people to absorb. Most farmers and ranchers find they are most comfortable doing whatever their neighbors are doing and thinking whatever their neighbors are thinking. This is known as "social harmony."

The idea that one would purposely go against the pre-existing "community of interest" as a business strategy is seen as akin to committing treason. And, the traitor doing so is "shunned" by the community.

Peter Drucker said that *it is this act of purposefully driving the entrepreneurial members of their community away that keeps rural areas poor.* Many entrepreneurs report the "liberation" of being ignored by their neighbors when they move to the city rather than facing outright hostility.

Interestingly, the movement of entrepreneurial urbanites back to the country may similarly "liberate" moribund rural economies.

France has probably taken the idea of "artisan" food production farther than any other developed country. They have done this because they largely agree with Stewart.

I read a report from France in which the French Agricultural Ministry admitted that their fight for the retention of high farm subsidies was a rear-guard action designed to delay the inevitable. The "inevitable" in this case being the loss of commodity agriculture in France to lower-priced foreign imports from the underdeveloped world.

Their great hope is to build a rural economy based upon high-value artisan food production. This economy was described as similar to a four-spoked wheel.

The four spokes of the wheel were:
- Artisan foods
- Landscape
- Local craft production
- Agritourism.

Interestingly, *the primary product of artisan food production was not considered to be the food but the beautiful landscape* its production created.

Thanks to the high-value products, the farms could be

small and diversified. This diversity and intensity would create the jumbled tapestry of small fields interspersed into a largely pastoral, largely open landscape with small, non-threatening patches of forest.

It is this attractive landscape, the French believe, that will draw urban tourists to the countryside. There, they will buy the artisan food production and non-industrial, hand-crafted items.

The key element for the rural economy *is the creation of a marketing interface* so that the primary producer of both the foods and the crafts would be able to sell his production direct to the consumer and thereby retain all of its profits. While urban farmers' markets bring the countryside to the urban consumers, the French want to bring their urban consumers to the countryside.

These urban visitors would spend the weekend in small inns and hotels and keep the local farmhouse restaurants full. Other rural entertainments such as horseback riding, canoeing and rafting would be provided. While it sounds Disneyesque, it was an idea that the urban French taxpayer liked and was willing to pay for. The only people who opposed this idea were — guess who? — the French farmers.

The idea of having to interact with urban consumers left many French farmers cold. They had made a lot of capital and mental investments in the industrial farming paradigm. Like many industrial farmers, they liked what they were doing, they just didn't like the pay.

As a protest to the Paris Ag Show, which celebrates artisanal agriculture and attracts some 650,000 urban attendees, these farmers started their own show to celebrate industrial agriculture at the same time on the other side of town.

> **All profit lies in the gap between what you know and what most people know.**
> **Thomas A. Stewart, Harvard Business School**

Whereas the Paris Ag Show was created to give Parisians a "taste" of the countryside through extensive food sampling, the industrial show was designed to celebrate the latest in tractors and machinery.

At first, the French officials thought this disaffection among the rural peasantry was going to be a major stumbling block to artisanal agriculture but it has subsequently proven not to have been. The reason for this was that rural France was being re-populated by urbanites seeking André Voisin's "pleasant life in the country." André Voisin was an early promoter of Management-intensive Grazing.

Today, half of all French farmers have lived in the cities at one time or another in their lives. All of these new people are "outsiders" to rural society and so do not care what their neighbors think about them. These new "urban" farmers have picked up the artisan food idea and are running with it.

While the vision of an artisan economy has not been completely fulfilled in France, artisan foods and agritourism are growing rapidly, and industrial ag is shrinking, so it may just be a matter of time. No doubt the continuing cutbacks in European commodity subsidies will speed this transition along.

Switzerland is a major believer in the concept of landscape as the primary tourist draw and both supports and subsidizes artisanal production.

The UK is gearing more of its farm subsidies into "landscape" items such as rebuilding stone walls and replanting hedgerows rather than supporting commodity production.

Due to its lengthier vacations, Europe's leisure industry is much more highly developed than that in North America and so

---

**What creates wealth is a source of unfair competitive advantage. The only two sources of this are:**
- **Differentiation**
- **Least cost production**
    **Thomas A. Stewart, Harvard Business School**

direct comparisons are difficult. Still there is enough evidence in North America that artisanal foods will become a sizable industry in the future.

Two things North America lacks are the French government's artisanal vision and its willingness to help build protective regulatory walls around its artisan producers.

Thomas Stewart in *The Wealth of Knowledge* said that some people hit the wall. Some people spend a lot of time and energy trying to dig under the wall. And some people build the wall.

He said ***all profitable producers try to build walls to limit competition*** either through rules and regulations, patents and proprietary processes.

"Free markets destroy profits. Always try to set up shop in a market that is less than free," he said.

In France, knock-off industrial products that look similar to artisanal food products have become the biggest headache to that country's artisanal producers. And, this is in a country that protects special labeling and distinctive containers for artisanal food products.

In North America, artisanal producers will have to find their political clout by allying with their urban consumers rather than with the traditional farm lobbies. These lobbies are there to protect the status quo and artisanal producers are seen as market usurpers and outsiders by traditional producers.

Stewart said there are three elements to a successful strategy in this highly competitive deflationary world:

- Have a unique value proposition.
- Have a source of control over your competition.
- Have a profit model that works.

He said all three of the elements could be found in a product that was unique and difficult to produce.

"The more differentiated your offering the more you can charge for it," he said.

"You can usually beat the price game by increasing the knowledge intensity of the product."

In other words, look for a product that requires a skill that takes time to learn. Like gourmet grass-finished beef.

The key word here is time. *Most people are willing to spend anything to get what they want except the time it takes to learn how to do it exceptionally well.*

Stewart summed up this necessity for a time element "wall" perfectly.

"Information is ever cheaper and knowledge ever more valuable.

"Knowledge involves expertise.

"Achieving it involves time.

"It endures longer than information.

"Sometimes forever."

## The Artisanal Alternative

Many farmers and ranchers have drifted into an industrial agricultural model without realizing what they were buying into. Harvard business professor, Michael Porter, said at the heart of any industrial activity is the substitution of capital for labor. *If the only labor unit on your ranch is you, this is definitely not a policy you want to follow.*

Industrialization lowers labor costs in two ways. One, by the outright substitution of machines for human labor, and two, by breaking skilled labor down into a series of simple, repetitive tasks capable of being done by unskilled labor.

For example, there is a lot of skill required to take a beef carcass apart if one person does it. However if 500 people are taught to make one cut each, it becomes unskilled work.

In the industrial model, all production methods must be scalable and subject to the economies of scale. Typically, in an industrial model *costs fall by 20% for every doubling of production.* This is why there is a race to be the biggest.

Publicly held companies have access to low-cost capital through stock sales. Their primary task is to create a competitive return for this capital. As long as this return is greater than they can get on a risk-free government bond, they will invest.

However, the acceptable rate of return for an investment is far too low for a small volume business or a startup. Small scale businesses need returns in the 30% to 150% range in order to grow from their own retained earnings.

Plentiful, low cost capital and industrialization are parts of the same whole. If you don't have the one you can't afford the other.

However, overworked farmers and ranchers saw the substitution of labor with capital as a way they could buy themselves increased leisure time. However, a *capital investment is only cost-effective if it allows you to do more work, not less.*

No industrialist is investing in a machine so that his workers can take a two hour lunch break rather than an hour.

In cropping agriculture, farmers also ignored the fact that their hugely expensive tractors, combines and attachments would only be doing useful work a few weeks a year. Yes, they allowed you to cover more acres in a day but they didn't lengthen the growing season by one minute.

The return to capital was so poor that land-based industrial agriculture suffered a total financial collapse in the early 1980s. The lie of this financial model can be seen in that it can only survive with massive government subsidies.

## The Factory Farm

Seeing the folly in that semi-industrial model, the smart boys in animal agriculture decided to sever all ties with the land and with climatic seasonality.

Following the pure industrial model, their "farm" would become a factory that just added value to purchased cheap commodity inputs. They would not grow any crops at all.

Thanks to falling rail freight rates, livestock operations could be located far from where the corn was actually grown. To avoid political problems and public scrutiny, most of these factories were located in remote Western locations or economically depressed areas in the South.

Preaching the religion of productivity, these guys were even able to get their suppliers to fund much of their capital investment.

Industrial feedyards were able to grow rapidly because they were able to talk other people into taking all of the heavy financial risk of cattle ownership. Their financial model was a hotel with a feedmill as a kitchen.

In a stroke of pure genius, the poultry and pork industrialists in collusion with rural Southern bankers got their "growers" to take on the huge land and capital risks of building the necessary production grow-out infrastructure they needed.

With debt up to their eyeballs for a facility with no other useful purpose, the "integrators" could tell the contract grower to jump and the growers had to jump or go bankrupt.

Like the supermarkets, these integrators used the power of their interposition between the growers and the consumer to milk them for extremely low cost capital.

In all of these examples, the farmers and ranchers voluntarily entered into these agreements to finance other people's dreams rather than their own.

All profit lies in the gap between what you know and what other people know. If you are naive about finance, you will probably be creating more wealth for other people than for yourself.

Keep in mind that ***productivity and profit are two entirely different concepts.*** Productivity is a measurement of physical inputs versus physical output.

In other words, productivity is about measuring things. Pounds of grain in versus pounds of beef, poultry or pork out. That accountants frequently dollar-ize these inputs does not negate the fact that ***productivity is about things and not money.***

What is often overlooked about productivity is that you can be just as productive doing the wrong thing as the right thing. A highly productive production model is not necessarily the most profitable one.

***The formula for profit is margin times volume less expenses.***

When looking at profit the amount of margin per unit of production and your expenses determines how much volume you must have. You can have a low volume of sales and still be profitable as long as your margin per unit of production is wide and your expenses are kept low.

The most profitable business is one that is small but growing fast. However, once the business outgrows the management ability of the founder, overhead costs skyrocket because hired management is very expensive. This is why some people purposely choose very small markets to serve.

***It is far better to work to make your business financially healthy that to try to make it grow. Healthy businesses naturally grow because they are doing the right things.***

I hate to admit it but the industrial animal factory model is a very viable production system. It takes trainloads of corn and stamps out great gobs of bland meat products at a very low cost. However, its capital intensity limits it as a model to those with access to lots of cheap capital.

And so, what about the rest of us? Is there no hope for those of us seeking to maximize the return to our own labor and land? Yes, there is.

There is an economic law called "The Law of Paradox" that states that the exact opposite of any viable economic system will also be economically viable.

Therefore, what we need to do is to turn the industrial model on its head and try to create its opposite.

## Artisanal Agriculture

■      Whereas the industrial model seeks to eliminate labor and particularly skilled labor, ***our model must be based upon retaining a significant component of skilled labor.***

■      Whereas the industrial model seeks to maximize the use of capital, ***ours must minimize it.***

■      Whereas the industrial model allows the price of the

output product to determine what it will pay for inputs, *our input prices must determine our output price.*

■       Whereas the industrial model seeks to deny the seasonality of nature, *ours must celebrate and embrace it.*

■       Whereas the industrial model seeks intentional blandness, *ours must seek exceptional flavor and uniqueness.*

■       Whereas the industrial model seeks to dominate distribution on a national and international scale, *ours must be directed to local and regional markets.*

■       Whereas the industrial model seeks attention with multimillion dollar television ads at the Super Bowl, *ours must be centered on word-of-mouth and customer referral.*

■       Whereas the industrial model runs on huge inputs of carbon-based energy, *ours must run largely on free solar energy.*

■       Whereas the industrial model sees time as money, *ours must see money as money.*

For lack of a better term, I call this mirror image of industrial agriculture Artisanal Agriculture.

The biggest problem with alternative agriculture today is that it seeks to incorporate bits and pieces of the industrial model and bits and pieces of the artisanal model. *This will not work.*

As Porter said *you must be one or the other*. If you are in the middle, you get the costs of the industrial model and the seasonal limitations and product variability of the artisanal product.

In other words, in the middle you get the worst of both worlds and not the best.

You can see the examples of this middle-of-the-road approach all around.

Most organic livestock production is just a small industrial model with higher cost inputs. Despite a premium price, the results have been as financially devastating as any other small scale industrial-based production model.

Industrial models will not scale down in size. They can

only scale up. Once you decide to move feed, the guy who can afford the biggest shovel is going to win.

Now, here's why you want to get out of the commodity business.

If you do not have at least some direct access to the consumer you will always be working for the person who does rather than yourself. He will tell you what and when you can produce, the price you will receive, and the recommended production model will be designed to maximize his return — not yours.

Remember, all of the power in a value chain lies with the person who creates the ultimate paying customer. *If you want to determine your own destiny, you will have to create your own customers.*

## Marketing

All businesses are made up of three distinct skills. These are *production, marketing and finance.* This trio has been described as similar to a three-legged dairy stool. If one leg is missing, the stool falls over.

The problem with all self-employed people is that no one person ever becomes really good at all three. Most of us choose the one we feel the most comfortable with and rely on others to supply the other two legs.

Personally, I like marketing the best. My idea of a fun day is to curl up with a really insightful marketing book.

In farming and ranching most of us like production and *there are times when production skills can pay exceptionally well. The grassfed beef industry is in just such a time.*

According to Harvard professor, Michael Porter, what determines which skill is most valuable is where the your industry is in its lifecycle. This is covered in Porter's rather difficult to read textbook *Competitive Strategy.*

He said most industries have four distinct phases in their lifecycle. They have an introductory phase, a growth phase, a mature phase and a decline phase.

A high technology industry can go through this entire
lifecycle in less than five years but most take at least 20 years,
and a few 40 years.

Learning new things results in what is called the learning
curve. Normally this is shown as an upward sloping line but
psychologically feels like a downward slope for several years.

Despite its frustrations, Porter said it is far better to be
poor and at the start of the start of something new than rich and
fully invested in an industry at the beginning of the end. ***Knowledge is worth the most when only a few have it, and this
only happens in a new industry.***

Now, back to that three-legged skill stool.

Porter said that in the introductory phase of a new
industry, production is the most important skill.

In the world of which comes first, the chicken or the egg,
it is the presence of product that creates demand. People cannot
determine whether they want a new food product until they can
actually taste it.

I went to a meeting in Montgomery, Alabama, where
about 40 political, academic and media elites from that state
gathered to hear a presentation by musician/grazier, Teddy
Gentry, about grassfed beef, its health advantages and its possible positive economic impact upon the state. Having a celebrity
present, most of the people invited showed up.

The group sat in polite silence through Teddy's presentation with few questions and no overt enthusiasm. They had heard
all this before. Auburn University had been researching grassfed
beef off and on since the late 1970s to no effect.

Following Teddy's presentation on beef, the group was
served an excellently prepared grassfed steak dinner. After the
first bite of Teddy's genetically-selected-for-tenderness grassfed
steak, the room lit up like it had suddenly been plugged into a
nuclear power plant.

One bite dispelled a lifetime of what they thought they
knew about grassfed beef.

By the time they finished their dinner, they were convinced that the state of Alabama was indeed in on the ground floor of something new that could eventually be really big. That's the power of product. The proof is in the pudding and not in the skillful arrangement of the words in the recipe.

## Purple Cows

Seth Godin has written an excellent book on why *in any new industry a killer product takes precedence over marketing and finance.* This book is called *The Purple Cow* and is short and very easy to read.

He said most people drive by herds of cows every day and don't really see any of them because of their familiarity. However, everyone would notice a purple-colored cow.

He said *getting noticed is the heart of all marketing.* The only way a new product will get noticed in today's crowded marketplace is *if it is dramatically different.*

He said the problem for all new products today is that the majority of consumers are happy with what they have. They aren't looking for a replacement and they don't like adapting to anything new.

"Marketing is no longer about making a product attractive or interesting or pretty or funny after it's designed and built — it's about designing the thing to be worthy in the first place."

In other words, what the product **IS** is the secret to marketing success.

Godin said only truly unusual products have any hope of success with today's jaded consumers.

"The only hope you have is to sell to people who like change, who like new stuff, who are actively looking for what it is you sell."

He said *most products aren't successful because the people who sell them aren't comfortable with the outrageousness required to get noticed.*

This is why most farmers and ranchers would love to fob the marketing off on someone else — often by forming a cooperative.

**Co-ops**

There are two basic kinds of farm co-operatives.

One is the buying co-operative whereby farmers pool their orders in order to secure volume discounts and lower freight rates. This form of co-op, originally known as a "boxcar co-op," has been very successful and I highly recommend it.

The reason it is successful is that *it requires no particular expertise to buy something.* In the original "boxcar co-op," the County Agent held the farmers' money and then paid for the product and the freight. What could be simpler?

The other type of co-operative is the marketing co-op.

In this type of co-op farmers hire someone to market their production in order to get a higher than commodity price. This form of co-op has been as spectacularly unsuccessful as the first form has been successful. This is because a marketing co-op is as complex as a buying co-op is simple.

The biggest reason for the general failure of marketing co-ops is that they are predicated on an unworkable business model.

Producers form marketing co-ops because they want a higher price for their production. This higher price to the producer is the primary "profit" goal of the co-op. However, this flies in the face of economic reality. *No real business has as its primary goal to pay a premium price for its production inputs.*

True businesses are structured to maximize the return to management and they try to pay as low a price for its inputs as competitively possible.

While farmers and ranchers often feel victimized by this, they do exactly the same thing to their own input suppliers. Note, that the previously mentioned "boxcar co-op" was not formed to maximize payments to the farm supplier but to minimize them.

Because maximizing the sales price of their production is their primary goal, there is no loyalty to the co-op if it cannot pay a premium price over the commodity market.

This is known as the "incentive trap."

If supply loyalty is solely dependent upon price, the producer can demand an ever higher price.

Consider the power difference between being a producer with a steer gaining two pounds a day on grass and a co-operative abattoir with a dozen highly paid people needing something to do.

This is why true businesses typically rely upon contract or their own production for a good portion of their daily input needs. They cannot count on producers being willing to sell when they need them to sell in order to maintain plant and marketing chain throughput. However, a "captive supply" is considered a terrible idea by producers who want to retain the freedom to withhold product for the highest price possible.

In other words, *they want the co-op to be there when they need it but want to be able to ignore it if they can get a better price elsewhere.* As a result of this lack of guaranteed supply, co-ops tend to form during periods of low commodity prices and quickly go out of business during periods of high commodity prices.

Another basic problem is that new co-ops tend to be formed around democratic ideals whereby everyone's input is considered equally important. However, if the management maxim is to do nothing until we all agree, nothing will get done.

Successful businesses are always dictatorships. Hopefully benevolent dictatorships, but dictatorships nevertheless. This means everyone in a business must submit and follow one person's vision.

As Ralph Waldo Emerson said, "All institutions are the lengthened shadow of one man."

Being independent folks, farmers and ranchers have a real problem with this.

As my Dad said, "There is no problem with the lack of leadership in farming and ranching. The problem is in the lack of followership."

## Consumer Buying Co-ops

Virginia grazier, Joel Salatin, has created a unique prototype for a consumer buying co-op. He calls these Metropolitan Buying Clubs (MBCs). As of summer 2005, he has seven of these in operation and several more in the process of forming.

MBCs are groups of consumers who are willing to pool their orders to make it worth Joel's time to deliver it to their neighborhood.

Joel's farm is approximately 150 miles south of the Washington, D.C., metroplex. He has his own on-farm store where he prefers to do his consumer marketing. This is because he has found that transportation costs are his biggest profit killer. "Transportation absolutely must carry its own weight," he said.

To accomplish this, he adds a 20 cents per pound transportation charge to his on-farm price. This rewards his on-farm store customers with lower prices and enables him to categorize transportation costs and track his true profitability.

Here are Joel's rules for consumers:

■      All orders must be received by 9 pm two days before delivery.

■      Each drop must average $1000 per delivery. New clubs have one year to achieve that level of sales.

■      If an established buying club drops below the minimum, Joel recommends that you cut it off. This fear of being cut off stimulates evangelistic marketing.

■      No membership costs. If you buy something, you've just joined.

■      A minimum of two purchases per year insures active, good-standing membership.

■      All communications should be electronically. Phone, fax or e-mail give quick turn arounds. Mail is too slow.

Here are Joel's rules for his delivery staff:

■      Be on time. These are busy mothers and have virtually no patience.

■ You must always send two people. One person should play the talker-schmoozer role and the other the hefter-bouncer. Joel said patrons love to talk to the farmer but somebody has to move material and watch the clock to stay on time. Again, you must have at least two people.

Here's Joel's advice on starting an MBC:

■ Use any connection you have. Friend, relative, business acquaintance.

■ Cold call on naturopaths, chiropractors, homeopaths, acupuncturists, wellness center director or food editor of your local paper.

■ To grow his MBC's he gives $10 off the next order for anyone who sends him a new customer who actually buys something.

■ He also takes good care of his MBC hostesses who allow him to use their home. He gives her freebies, gifts and thank you notes.

■ He makes sure every member of the MBC knows the gross sales volume of every other MBC after each visit. This stimulates lagging MBC's and rewards winning MBC's.

■ All patrons must deal directly with the farm. This is incredibly efficient.

■ Joel services each club on about a six to eight week basis. It is much more efficient to handle fewer, larger invoices than the opposite. Most customers spend about $250.

■ Everything is delivered frozen. Period. Too much can go wrong with fresh product.

"The beauty of an MBC is that customer communication and product assimilation can be done on my schedule, not someone else's. Farmgate sales are wonderful, but they interrupt our work during prime farmwork time," Joel said.

"Metropolitan Buying Clubs can bridge the gap between the farmers' markets and supermarket sales."

*There are a lot more instances of successful businesses farming out production than there are of successful businesses farming out marketing.*

But, again, at this stage of development in grassfed beef the product is the marketing, so there is no conflict.

Seth Godin warns that a goal of making it "as good as" the competition won't work.

*"Don't try to make a product for everybody, because that is a product for nobody.* The everybody products are all taken.

"How can you market yourself as more bland than the leading brand? The real growth comes from products that annoy, offend, are too expensive, or too complicated to produce or too something."

He said a niche market product would always be too "too" for some people but just perfect for others.

Ironically, Godin said *the narrower and more focused a product starts the greater is its chance of achieving large scale success later on.*

"The way you break through to the mainstream is to target a niche instead of a huge market. You want to create a product that is so focused that it overwhelms that small slice of the market that really and truly will respond to what you sell.

"After it dominates the original niche, if you're good and you're lucky, that product will diffuse. After it dominates the original niche, it will migrate to the masses."

Because no one can be sure of the size of a new market, initial production has to be small and tentative. However, if this product attracts even a small group of enthusiastic consumers it can grow from its own cash flow because it can earn premium prices as long as supply is less than demand.

We are already far enough along to have this situation. Currently, grassfed beef sells for 10 to 30% more than commodity grainfed beef.

## How big is the ultimate market for gourmet grassfed beef?

Harvard professor, Michael Porter said that the *first criteria of a new market pioneer was the willingness to take on risk.*

The reason for this is that *the size of all new markets are not only unknown but unknowable.*

There is no way to market research a product that cannot be fully sampled. This means that the product must be produced on a sizable scale so that consumer reaction to it can be accurately judged.

However, even if the consumer appears delighted with the product and wants to buy more, new industries face constraints that keep them smaller than their market until they can be overcome. And, this lag period is guaranteed to create frustration among excited marketers.

I have taken Porter's list of growth constraints and have modified them to make them more relevant to the grassfed meats industry.

### 1. Difficulty in Finding Raw Materials

Due to the biological lag in cattle, supply cannot be quickly increased even if the market is clamoring for it. Most early market pioneers will not start a second set of cattle until they have sold the first set. In a 24-month program, this produces a considerable lag in ramping production up to satisfy new market needs.

### 2. Prices for Available Raw Materials Skyrocket in Price

Porter said that many marketers get excited about discovering a virgin market and want to "own it" before the competitors wize up. This desire for expansion at any cost produces a competitive scramble for the few available supplies of market ready cattle.

Quite often the wide margin found in pioneer markets is bid into the raw materials by overzealous and excited marketers. This not only sends a false signal to the producers of the raw materials but can result in the marketer going bankrupt at the same time the increase in supply finally appears.

This results in the "boom bust" pattern seen in the early stage of almost all new industries.

### 3. Absence of a Value Chain

All industries depend upon suppliers of products and services to make it possible. In the early stages, these do not exist and must be cobbled together or jury-rigged from other industries. The development of such value chain pieces as abattoirs, grass-friendly genetics suppliers and meat distributors must be encouraged and taught the unique needs of the new industry.

### 4. Shortage of Skilled Laborers

In a new industry there is always a dire shortage of people trained in the nuances of the production paradigm. This greatly slows growth because the founder of the embryonic company cannot turn over the lesser duties because there is no one available for hire trained in them.

In the early stages of an industry, employees who do learn these specialized skills quite often start their own companies rather than hiring out to capitalists wanting in on the new industry but lacking the necessary new skills.

### 5. Absence of Product Standards

Inability of the pioneers to agree on product, technical and genetic standards accentuates problems in the supply of raw materials and impedes cost improvements.

### 6. Customer Confusion Over Product Variations

The lack of a consistent standard of what the product is and how it is produced creates great confusion among consumers. Many who find that what they thought was a true grassfed product get angry when they find out it is not and believe the whole industry is fraudulent.

### 7. Erratic Product Quality

Erratic product quality, even if from only a few firms, can negatively impact the image and credibility of the whole industry. High quality producers must quickly differentiate themselves by winning contests and obtaining objective third party vetting from the food media and critics.

Standards and protocols should be written to encourage low quality producers to either improve or leave the industry.

## 8. Poor Image and Credibility with Financial Community

The boom-bust cycle of a new industry inevitably results in early lenders getting burned. This makes lending very difficult for all participants after an over-supply bust.

## 9. High Costs Due to Low Volumes

New industries are forced to pay high job shop rates for production services such as abattoir work because of low and erratic volumes. These high costs force the pioneer product to seek premium-priced market niches to maintain margins.

## 10. The Government Becomes Interested

Once a new industry becomes visible to the consumer, it also becomes visible to government bureaucrats who discover it has few rules and regulations they would like to put on it.

This government interest is often stoked by threatened establishment industries who seek to use government regulations to slow the growth of the new industry.

This is why new industries cannot ignore politics and must early on seek out politicians willing to protect and advocate it. These politicians must be rewarded with favorable publicity and campaign contributions.

The point here is that no new industry can be born and skip childhood and adolescence. All have to go through the awkwardness of learning how to function as a grown up.

If you are a pioneer grassfed marketer and are feeling frustrated at how difficult and slow things seem to be, what you are feeling is normal and is not peculiar to the grassfed industry. *Don't try to skip the problems. Recognize they exist and plan to grow through them.*

# CHAPTER 15

# Ready? Set? Go for It!

This book has taken me several years to write. When I started writing it, I had know idea how fast the grassfed beef industry in America would develop.

As I write these words in the early summer of 2005, another Mad Cow scare has pushed grass-finished steers in Oklahoma to $1.35 a pound liveweight. This compares with 72 cents for grain-finished steers, which have fallen 12 cents a pound in price in one week. While this huge price gap is unusually wide it illustrates a couple of things. One — the demand for grass-finished beeves is far in excess of supply, and two — the more consumers know about the details of grain-finished beef the less they like it. Both of these are positive for the continued growth of the grass-finished industry.

Since 1999, the grassfed industry has grown from less than 40 individual marketers of grassfed meats to well over 1000 active marketers — that's in just five years. Many of these marketers access beeves from a dozen or more producers, so a sizable industry is developing that is still below the radar screen of most commodity analysts.

Currently, *The Stockman GrassFarmer Magazine*, which serves as the trade journal for the pasture-based food industry, has over 11,000 paid subscribers and is thriving. This nascent industry is rapidly developing its own seedstock, inputs and specialized consultants. Internet sales are booming. The early beginnings of a national retail distribution network are being cobbled together. As product seasonality declines, the current constraint of the lack of abattoirs will naturally take care of itself.

In summer 2005, for the first time in 20 years, I was able to go into a supermarket in New Orleans and buy a beautifully marbled steak *proudly* labeled "grassfed." This steak was priced at a mere 300% premium in price to the commodity grainfed product and yet the store's butcher told me he couldn't keep it in stock.

And we've only just begun!

In another 20 years, we will look back fondly on the next few years and remember them as grass-finished meats' "Golden Era." Only in a pioneering industry does the producer have any power. This is because the product *is* the market. Consumers are drawn to it because of its differences *from* grain-finished beef. Thanks to its inherent health attributes, the consumer has a product to enjoy without guilt.

As Dr. Tilak Dhiman of Utah State University has pointed out, we have a product that is provably far more healthy than the mainstream product that is possible with organic foods. Therefore, it is reasonable to assume that the grassfed industry will have an even steeper growth curve than the 20% per annum increase organic has had.

Of course, you can't sell what doesn't exist. Therefore, the growth rate of the industry will primarily be determined by the number of graziers who convert from selling feeder cattle to the grain-finished commodity industry to selling grass-finished beef. The current $600 or more per head profit premium of the grassfed animal should help to capture producers' attention.

The key is to keep focused on the eating quality of the product as the consumer defines quality — tenderness and taste. This book has shown you that a high eating quality grass-fed product *can* be produced year around in much of the USA.

New industries create new wealth for new people. I hope you will decide to be one of them. Go for it!

## Cattle Futures?

"For several years now, an alternative, post-industrial food chain has been taking shape, its growth fueled by one "food scare" after another. Alar, G.M.O.s, rBGH, E. coli 0157:H7, now B.S.E.

Whatever science told us about the risks of these novel industrial entrees and sides, something else told us we might want to order something more appetizing: organic, hormone-free, grass-finished.

It might cost more, but it's possible again to eat meat from a short, legible food chain consisting of little more than sunlight, grass and ruminants.

Back to the future: a 21st century savanna.

If, as seems probable, this landscape should now expand at the expense of the feedlot, then something good - even beautiful — will have come of this poor mad cow."

Michael Pollan,
January 11, 2004 issue
*The New York Times Magazine*

# INDEX

# More from Green Park Press

**AL'S OBS, 20 Questions & Their Answers** by Allan Nation. By popular demand this collection of Al's Obs is presented in question format. Each chapter was selected for its timeless message. 218 pages. **$22.00\***

**COMEBACK FARMS, Rejuvenating soils, pastures and profits with livestock grazing management** by Greg Judy. Takes up where *No Risk Ranching* left off. Expands the grazing concept on leased land with sheep, goats, and pigs in addition to cattle. Covers High Density Grazing, fencing gear and systems, grass-genetic cattle, developing parasite-resistant sheep. 280 pages. **$29.00\***

**GRASSFED TO FINISH, A production guide to Gourmet Grass-finished Beef** by Allan Nation. How to create a forage chain of grasses and legumes to keep things growing year-around. A gourmet product is not only possible year around, but can be produced virtually everywhere in North America. 304 pages. **$33.00\***

**KICK THE HAY HABIT, A practical guide to year-around grazing** by Jim Gerrish. How to eliminate hay - the most costly expense in your operation - no matter where you live in North America. 224 pages. **$27.00\*** or Audio version - 6 CDs with charts & figures. **$43.00**

**KNOWLEDGE RICH RANCHING** by Allan Nation. In today's market knowledge separates the rich from the rest. It reveals the secrets of high profit grass farms and ranches, and explains family and business structures for today's and future generations. The first to cover business management principles of grass farming and ranching. Anyone who has profit as their goal will benefit from this book. 336 pages. **$32.00\***

**LAND, LIVESTOCK & LIFE, A grazier's guide to finance** by Allan Nation. Shows how to separate land from a livestock business, make money on leased land by custom grazing, and how to create a quality lifestyle on the farm. 224 pages. **$25.00\***

**MANAGEMENT-INTENSIVE GRAZING, The Grassroots of Grass Farming** by Jim Gerrish. Details MiG grazing basics: why pastures should be divided into paddocks, how to tap into the power of stock density, extending the grazing season with annual forages and more. Chapter summaries include tips for putting each lesson to work. 320 pages. **$31.00\***

# More from Green Park Press

**MARKETING GRASSFED PRODUCTS PROFITABLY** by Carolyn Nation. From farmers' markets to farm stores and beyond, how to market grassfed meats and milk products successfully. Covers pricing, marketing plans, buyers' clubs, tips for working with men and women customers, and how to capitalize on public relations without investing in advertising. 368 pages. **$28.50**

**NO RISK RANCHING, Custom Grazing on Leased Land** by Greg Judy. Based on first-hand experience, Judy explains how by custom grazing on leased land he was able to pay for his entire farm and home loan within three years. 240 pages. **$28.00***

**PADDOCK SHIFT, Revised Edition Drawn from Al's Obs, Changing Views on Grassland Farming** by Allan Nation. A collection of timeless Al's Obs. 176 pages. **$20.00***

**PASTURE PROFIT$ WITH STOCKER CATTLE** by Allan Nation. Profiles Gordon Hazard, who accumulated and stocked a 3000-acre grass farm solely from retained stocker profits and no bank leverage. Nation backs his economic theories with real life budgets, including one showing investors how to double their money in a year by investing in stocker cattle. 192 pages **$24.95*** or Abridged audio 6 CDs. **$40.00**

**QUALTIY PASTURE, How to create it, manage it, and profit from it** by Allan Nation. No nonsense tips to boost profits with livestock. How to build pasture from the soil up. How to match pasture quality to livestock class and stocking rates for seasonal dairying, beef production, and multispecies grazing. Examples of real people making real profits. 288 pages. **$32.50***

**THE MOVING FEAST, A cultural history of the heritage foods of Southeast Mississippi** by Allan Nation. How using the organic techniques from 150 years ago for food crops, trees and livestock can be produced in the South today. 140 pages. **$20.00***

**THE USE OF STORED FORAGES WITH STOCKER AND GRASS-FINISHED CATTLE.** by Anibal Pordomingo. Helps determine when and how to feed stored forages. 58 pages. **$18.00***

* All books softcover. Prices do not include shipping & handling

## To order call 1-800-748-9808
## or visit www.stockmangrassfarmer.com

# Questions
# about grazing ???????
# Answers *Free!*

While supplies last, you can receive a Sample issue designed
to answer many of your questions. Topics include:
* How You Can Beat the High Cost of Cow Depreciation
* What Is Your Livestock Business' Breeding Objective?
* Selling Grassfed Beef on Flavor and Production Practices
* Three Proven Prototypes for Pastured Poultry
* Recycle Nutrients
* High Stock Density Grazing and Daily Shifts
* Underappreciated Weeds
* The Affects of Soil Acidity on Grazing Animals
* And more

Green Park Press books and the Stockman Grass Farmer
magazine are devoted solely to the art and science of turning pastureland into profits through the use of animals as
nature's harvesters. To order a free sample copy of the
magazine or to purchase other Green Park Press titles:

P.O. Box 2300, Ridgeland, MS 39158-2300
1-800-748-9808/601-853-1861

Visit our website: www.stockmangrassfarmer.com
E-mail: sgfsample@aol.com

Name _____

Address _____

City _____

State/Province_____Zip/Postal Code _____

Phone _____

| Quantity | Title | Price Each | Sub Total |
|---|---|---|---|
| ____ | 20 Questions (weight 1 lb) | $22.00 | _____ |
| ____ | Comeback Farms (weight 1 lb) | $29.00 | _____ |
| ____ | Grassfed to Finish (weight 1 lb) | $33.00 | _____ |
| ____ | Kick the Hay Habit (weight 1 lb) | $27.00 | _____ |
| ____ | Kick the Hay Habit Audio - 6 CDs | $43.00 | _____ |
| ____ | Knowledge Rich Ranching (wt 1½ lb) | $32.00 | _____ |
| ____ | Land, Livestock & Life (weight 1 lb) | $25.00 | _____ |
| ____ | Management-intensive Grazing (wt 1 lb) | $31.00 | _____ |
| ____ | Marketing Grassfed Products Profitably (1½) | $28.50 | _____ |
| ____ | No Risk Ranching (weight 1 lb) | $28.00 | _____ |
| ____ | Paddock Shift (weight 1 lb) | $20.00 | _____ |
| ____ | Pa$ture Profit$ with Stocker Cattle (1 lb) | $24.95 | _____ |
| ____ | Pa$ture Profit abridged Audio -- 6 CDs | $40.00 | _____ |
| ____ | Quality Pasture (weight 1 lb) | $32.50 | _____ |
| ____ | The Moving Feast (weight 1 lb) | $20.00 | _____ |
| ____ | The Use of Stored Forages (weight 1/2 lb) | $18.00 | _____ |
| ____ | Free Sample Copy *Stockman Grass Farmer* magazine | | _____ |

Sub Total _____

Mississippi residents add 7% Sales Tax _____

Postage & handling _____

TOTAL _____

Foreign Postage:
Add 40% of order

**We ship 4 lbs per package maximum outside USA.**

| Shipping | Amount | Canada |
|---|---|---|
| Under 2 lbs | $5.60 | |
| 2-3 lbs | $7.00 | 1 book $13.00 |
| 3-4 lbs | $8.00 | 2 books $20.00 |
| 4-5 lbs | $9.60 | 3 to 4 books $25.00 |
| 5-6 lbs | $11.50 | |
| 6-8 lbs | $15.25 | |
| 8-10 lbs | $18.50 | |

**www.stockmangrassfarmer.com**

*Please make checks payable to*

**Stockman Grass Farmer
PO Box 2300
Ridgeland, MS 39158-2300**

**1-800-748-9808
or 601-853-1861
FAX 601-853-8087**

Name _____

Address _____

City _____

State/Province_____Zip/Postal Code _____

Phone _____

| Quantity | Title | Price Each | Sub Total |
|---|---|---|---|
| ____ | **20 Questions** (weight 1 lb) | **$22.00** | _____ |
| ____ | **Comeback Farms** (weight 1 lb) | **$29.00** | _____ |
| ____ | **Grassfed to Finish** (weight 1 lb) | **$33.00** | _____ |
| ____ | **Kick the Hay Habit** (weight 1 lb) | **$27.00** | _____ |
| ____ | **Kick the Hay Habit Audio - 6 CDs** | **$43.00** | _____ |
| ____ | **Knowledge Rich Ranching** (wt 1½ lb) | **$32.00** | _____ |
| ____ | **Land, Livestock & Life** (weight 1 lb) | **$25.00** | _____ |
| ____ | **Management-intensive Grazing** (wt 1 lb) | **$31.00** | _____ |
| ____ | **Marketing Grassfed Products Profitably** (1½) | **$28.50** | _____ |
| ____ | **No Risk Ranching** (weight 1 lb) | **$28.00** | _____ |
| ____ | **Paddock Shift** (weight 1 lb) | **$20.00** | _____ |
| ____ | **Pa$ture Profit$ with Stocker Cattle** (1 lb) | **$24.95** | _____ |
| ____ | **Pa$ture Profit abridged Audio -- 6 CDs** | **$40.00** | _____ |
| ____ | **Quality Pasture** (weight 1 lb) | **$32.50** | _____ |
| ____ | **The Moving Feast** (weight 1 lb) | **$20.00** | _____ |
| ____ | **The Use of Stored Forages (weight 1/2 lb)** | **$18.00** | _____ |
| ____ | Free Sample Copy ***Stockman Grass Farmer*** magazine | | _____ |

Sub Total _____

Mississippi residents add 7% Sales Tax _____

Postage & handling _____

TOTAL _____

| Shipping | Amount |
|---|---|
| Under 2 lbs | $5.60 |
| 2-3 lbs | $7.00 |
| 3-4 lbs | $8.00 |
| 4-5 lbs | $9.60 |
| 5-6 lbs | $11.50 |
| 6-8 lbs | $15.25 |
| 8-10 lbs | $18.50 |

Canada
1 book $13.00
2 books $20.00
3 to 4 books $25.00

Foreign Postage:
Add 40% of order

**We ship 4 lbs per package
maximum outside USA.**

www.stockmangrassfarmer.com

*Please make checks payable to*

**Stockman Grass Farmer
PO Box 2300
Ridgeland, MS 39158-2300**

**1-800-748-9808
or 601-853-1861
FAX 601-853-8087**